Citizens Band Radio Handbook

by

David E. Hicks

HOWARD W. SAMS & CO., INC.
THE BOBBS-MERRILL CO., INC.
INDIANAPOLIS · KANSAS CITY · NEW YORK

FOURTH EDITION

EIGHTH PRINTING—1975

International Standard Book Number: 0-672-20839-3
Library of Congress Catalog Card Number: 70-149396

Preface

The Citizens Radio Service has made two-way radiocommunications available to the general public for the first time. For many years, the use of this medium was restricted to those who could show that the intended operation was in the interest of public safety, necessity, or convenience. This permitted police and fire departments, public utilities, and similar agencies to obtain station authorizations. It also allowed those who could qualify for an amateur radio license to engage in two-way radio communications.

This new and revised fourth edition is intended to introduce you to all phases of Citizens Band Radio—the service that allows you, "John Q. Citizen," to utilize two-way radio for your own personal use. This book provides the answers to many problems and questions which may confront those who are now using, or intend to use, CB equipment. Included are facts and discussions concerning the Citizens Radio Service, how to obtain a CB license, operating procedures, and other related subjects of concern. Examples of the latest equipment and accessories are also included to guide you in selecting the equipment best suited for your specific application. For the more technically oriented reader, circuit analyses, troubleshooting hints, and servicing data are included.

It is hoped that this book, in some way, will provide a better understanding of the subject and contribute to your enjoyment and use of CB two-way radio.

DAVID E. HICKS

Contents

CHAPTER 5

CHAPTER 6

CHAPTER 7

CHAPTER 8

CHAPTER 9

APPENDIX I

APPENDIX II

APPENDIX III

1

The Citizens Radio Service

Radio has become one of the most widespread communications media of our times. However, until the introduction of the Citizens Radio Service, two-way radio was not available to the general public.

WHAT IS CITIZENS RADIO?

The Citizens Radio Service was established in 1947 by the Federal Communications Commission (FCC) to permit personal short-distance radiocommunications, signaling, and remote control by radio signals—all designed to fulfill a definite need in connection with both business and personal activities. This service made available to the general public, for the first time, the long-awaited convenience of a two-way radiotelephone system that could be used by practically anyone.

At this time, however, only two classes of license—Class A and Class B—were available. The frequencies allocated for both classes were within a band extending from 460 to 470 megahertz (MHz) in the ultrahigh frequency (uhf) region of the spectrum. Immediately below this band are frequencies assigned to remote pickup, industrial, land transportation, public safety, and domestic public services. Above 470 MHz are the uhf television channel assignments.

The Class-A and -B Citizens band (CB) stations were intended to provide two-way radiotelephone wherever the need arose. For example, a business could use CB radio for dispatching its delivery trucks, or an individual in his automobile could communicate with his home—possibly just to inform his wife that he was bringing company home for dinner.

Although its uses were practically unlimited, CB radio was rather slow in gaining acceptance. The two principal reasons for this were the cost of the equipment and the limited range of communications. Class-A stations could operate with much higher transmitter power than those with Class-B ratings, but the equipment was also much more complex and expensive. Conversely, although some of the Class-B equipment marketed was reasonably priced, there was still the problem of its limited range.

One of the many factors that determine the range of a transmitter is its operating frequency. At uhf frequencies, radio waves tend to travel in a straight line between transmitting and receiving antennas; they do not reflect and bend as much as the lower-frequency waves do and they are attenuated more rapidly. This is illustrated in Fig. 1-1; the uhf signal travels more or less in a line-of-sight path between the two antennas. Intervening terrain or objects such as trees or buildings greatly attenuate, and sometimes completely block the signal. This, of course, limits the usable communicating range at these frequencies. The lower-frequency signal is reflected more readily and reaches the receiving antenna with little loss. (This phenomenon is discussed at length in Chapter 4.)

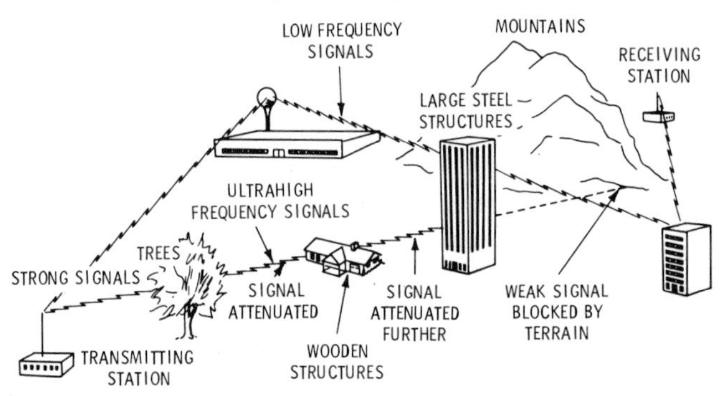

Fig. 1-1. An example of the uhf signal's line-of-sight characteristics.

Shortly after the Class-A and -B services were introduced, a third class was announced. Known as Class C, it was reserved for remote control of such devices as model airplanes, garage doors, etc. (radiotelephone communications are not permitted in this class).

Realizing the shortcomings of uhf communications and the need for practical Citizens-band radio, the FCC in 1958 allocated a number of channels for use by a new service labeled Class D. This is the class that accounts for the widespread popularity of CB radio today. The frequencies that were formerly assigned in the 11-meter amateur band were reallocated for the Class-D stations, and at these lower frequencies more reliable communications are obtained and over much greater distances. Furthermore, Class-D equipment is readily available in a variety of styles and at reasonable prices.

USES OF CB RADIO

CB radio uses are practically unlimited. Equipment can be permanently installed at a fixed location, in vehicles, or carried in the hand. In a car, CB radio can provide communications with your home or office; doctors may use it to communicate with the hospital as well. It can be used in boats and airplanes, on hunting, fishing, or camping trips, or even on a golf cart. Businesses are finding CB radio helpful. Much time and money can be saved with radio-dispatched service and delivery trucks—to say nothing of the more efficient service. Countless others are finding Citizens radio equally as helpful—radio and TV technicians, plumbers, police and fire departments, surveyors, and highway maintenance crews, to name just a few.

Each day new ways of utilizing this versatile medium are being discovered. More and more CB antennas are appearing atop hotels, motels, restaurants, service stations, and garages. You can make reservations at a motel or hotel or order your dinner while traveling along the highway. If you are stranded along the road with car trouble, help is no further away than your CB microphone. In many areas, local CB clubs have posted signs on the outskirts of the city listing the CB channels that are monitored. There is also a national *H*ighway *E*mergency *L*ocating *P*lan (HELP) in which thousands of miles of highway are monitored by rescue units, police, hospitals, auto-repair centers, and other CB'ers as an aid to motorists who are in need of assistance.

OBTAINING THE LICENSE

Since the CB transmitter radiates energy within the radio-frequency spectrum, it comes under the jurisdiction of the Federal Communications Commission. Before such a transmitter can be placed on the air, a station license must be obtained from this governmental agency. Unlike most other radio services, however, a CB license is fairly easy to obtain and no examination is required. However, the FCC has set up specific rules and regulations for the operation of CB equipment; failure to comply can result in a fine, and/or suspension or revocation of the station license. One of the limitations concerns the amount of transmitter power allowed, stated as so many watts of power input and power output. The actual rf output power of a transmitter is always less than the dc input power due to inherent rf circuit losses. With the exception of Class-A stations and operation on one Class-C channel, all CB transmitters are limited to an input power of 5 watts and an rf output of 4 watts. CB transmitters with a power input of 100 milliwatts or less require no license whatsoever. Most of the small hand-carried units fall into this category.

Who Is Eligible?

Any citizen of the United States, male or female and 18 years or older, is eligible for a Citizens-band license. Persons under 18 may operate CB radio equipment—but only under the direct supervision of a licensee who will assume full responsibility for proper station operation. Applicants for a Class-C Citizens-band license (used for remote-control) need only be 12 years of age or older to obtain a license.

Applicants are not required to learn code, nor to take any oral or written examination. No operator's license is required—only a station license authorizing the installation and operation of the equipment. (Equipment maintenance is subject to certain restrictions to be discussed later.)

A Citizens radio license will not be granted to aliens, nor to any person representing an alien. The same is true for a representative of a foreign government or to any corporation in which a director or officer is an alien. Specific eligibility requirements are defined in § 95.7 and 95.13 of the FCC Rules and Regulations contained in Part 95 of Volume VI.

Filing the Application

All applications for CB licenses are made on standard FCC forms. Form 400 is used for Class-A station licenses and Form 505 for Class-B, -C, and -D licenses. These forms are usually included with new CB gear; however, they can be obtained by writing the Washington, D.C. office of the Commission or the nearest FCC Field Engineering Office (listed in the Appendix).

One entry requires you to signify that you possess a *current* copy of Part 95 of the FCC *Rules and Regulations* governing the Citizens Radio Service. These are included in Volume VI, which can be purchased for $5.35 by writing the Superintendent of Documents, U. S. Government Printing Office, Washington, D.C. 20402. Part 95 of the rules and regulations is included in the Appendix of this book for your convenience.

All applicable questions on the form must be answered. Failure to do so may result in the application being marked "incomplete" and returned. One of the questions concerns the number of transmitters to be operated. Any number of transmitters can be covered by a single license as long as they are all under the control of the licensee and in the same geographical area; however, only one applicant (which can be a corporate group or company) can be licensed for the same equipment. The actual number of transmitters (the FCC is not concerned with receivers) must be listed on the application. If more transmitters are to be added after issuance of the license, an application for license modification must be submitted.

After all questions on the form have been answered, the applicant must sign it. Notarization is not necessary; however, making willful false statements is punishable by fine and imprisonment. The completed application for a Class-B, -C, or -D station license should be mailed to the FCC office at 334 York Street, Gettysburg, Pennsylvania, 17325. Class-A station applications, special requests, or correspondence relating to applications for any CB license should be mailed directly to the FCC office at Washington, D.C. 20554. Because of the many applications being received by the FCC, it is not unusual for processing to require several weeks.

Many persons purchase CB equipment before obtaining a license. Remember that the license must be in the applicant's possession before he can operate the equipment. The mere fact that an application for a license has been submitted does not constitute authorization to begin operation. Wait until you receive the license before you press that mike button.

License Fees

Except as provided in Subpart G, § 1.1105 of the FCC regulations, a fee of $4.00 must accompany the application for a new, renewal, or modification of a Class-A, -C, or -D license. A duplicate license can be obtained for a fee of $2.00.

Renewals and Modifications

All licenses are issued for a period of five years from the date of original issuance or renewal. Application for renewal of a CB license is made on the same form as for a new station license and should be filed at least 60 days before the license expires.

From time to time certain changes in equipment, location, etc., may be required. Should any such changes conflict with the terms on the current license, an application for modification must be submitted to the FCC. An application for license modification is made in the same manner as an application for a new station license. As in the case of renewals the same application form is employed. Upon receipt of the revised license, the applicant must forward the old one to the FCC. Refer to § 95.35 of the FCC Rules and Regulations for changes that require a license modification. Remember to include the fee when making an application for renewal or modification.

CB LICENSE CLASSES

Class-A License

The Class-A license permits the use of a much higher transmitter power than any other class, and also provides a variety of operating frequencies from which to choose. Moreover, the antenna height is not nearly as restricted as for Class-B and Class-D stations.

Frequency Allocations—The frequencies or frequency pairs listed in Table 1-1 are available for assignment to base and mobile stations operating in the Class-A Service, subject to the provisions of § 95.41(a) of the governing FCC regulations.

Table 1-1. Class-A Frequencies*

Base and Mobile		Mobile Only	
462.550	462.650	467.550	467.650
462.575	462.675	467.575	467.675
462.600	462.700	467.600	467.700
462.625	462.725	467.625	467.725

*All frequencies in MHz

Once a Class-A frequency, or pair of frequencies, is selected and assigned, prior FCC approval must be obtained before a change in frequency can be made. Such a change would also require modification of the existing station license. Fig. 1-2 shows a typical two-way radio that has been type-accepted for Class-A operation.

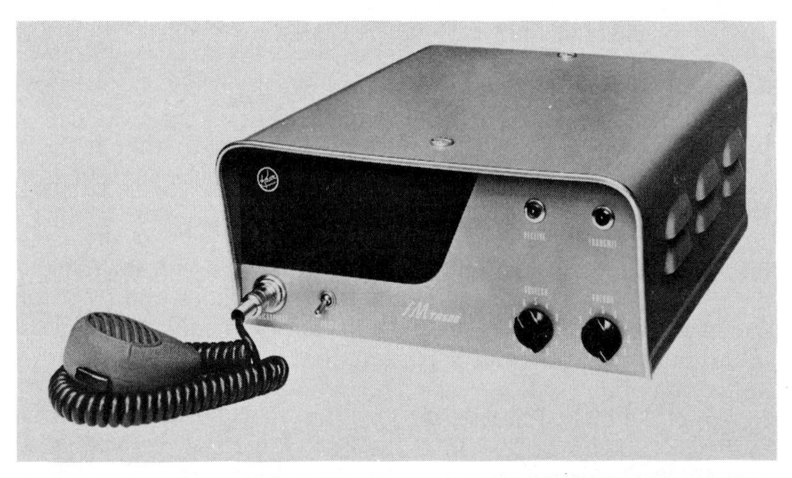

Courtesy Kaar Engineering Corp.

Fig. 1-2. A two-way radiotelephone transceiver. FCC-approved for Class-A operation.

Power Limitations—The maximum plate input power is 60 watts for a Class-A station (48 watts rf output). This is the highest permitted any class in the Citizens Radio Service.

Emission—Class-A units are normally authorized to transmit radiotelephone only, although the use of tone signals or signaling devices used solely to actuate receiver circuits (such as tone squelch) is permitted. The use of tones to attract attention is prohibited, however. Application for such authorization must be submitted to the FCC, stating not only the emission desired, but also why it is needed and what bandwidth will be required to provide satisfactory communications. Each Class-A license lists not only the type of emission authorized but also includes a prefix designating the maximum bandwidth it can occupy. Class-A stations transmit either a-m (amplitude modulation) or fm (frequency modulation) radiotelephone signals. The differences between the two, illustrated in Fig. 1-3, are as follows: In an a-m transmitter, the *frequency* of the carrier remains constant and its *amplitude* is changed in intensity, or modulated, by the voice signal (Fig. 1-3A). The amount of change is designated as the percentage of modulation.

In an fm transmitter, the *amplitude* of the rf carrier remains constant, and the *frequency* is varied above and below its normal value by the voice signal (Fig. 1-3B). As you can see, the names *amplitude* and *frequency* modulation are quite descriptive.

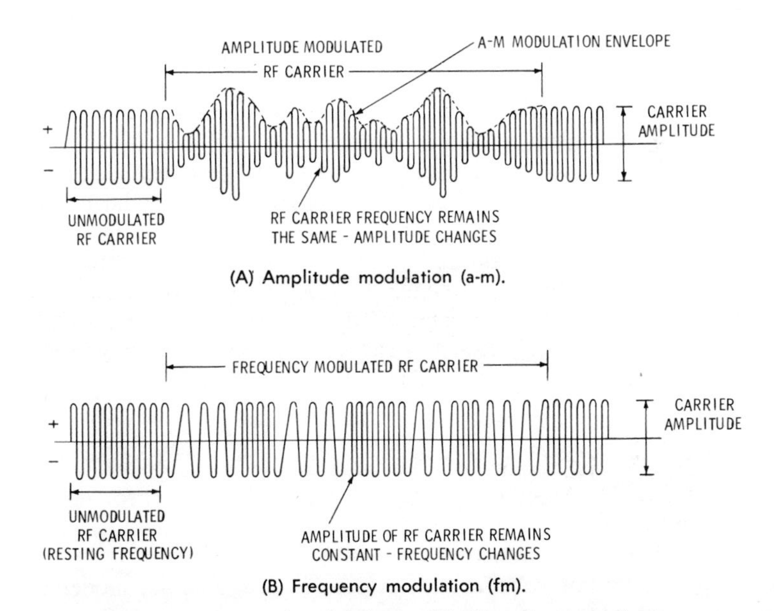

(A) Amplitude modulation (a-m).

(B) Frequency modulation (fm).

Fig. 1-3. Two types of modulation used in CB radio communications.

Class-B License

Shortly after Class-B operation was started, several manufacturers made equipment available at a price that appealed to almost everyone. This accounted for much of the popularity of Class-B instruments during the infancy of CB radio.

While the equipment used in Class-B operation was simpler, the maximum permissible power input was not nearly as high as for Class A.

The big problem of range confronted Class-B stations to a greater degree because of the power limitations. Many purchasers of Class-B equipment were disappointed at the results. Not understanding the problems at uhf frequencies, they expected too much. To add to the troubles of both the FCC and equipment manufacturers, unqualified persons often attempted to adjust or modify their equipment in hopes of improving performance. This usually resulted in off-frequency operation or a completely inoperative unit.

Frequency Allocations—Class-B stations are authorized to operate on 465 MHz only, unless the equipment has been type-accepted for Class-A operation. In this event, the station may also be operated on any of the Class-A frequencies. Equipment that is not crystal controlled must have a type-acceptance number which signifies that it meets all FCC technical requirements. Some manufacturers request type-acceptance for crystal-controlled equipment as well.

Power Limitations—Maximum plate power input for a Class-B station is 5 watts—quite a bit less than the 60 allowed a Class-A station. The actual output power of the transmitter is limited to 4 watts. However, it may be only 3 watts, or only a fraction of a watt, depending on the efficiency of the transmitter.

Emission—Class-B stations are permitted to use either a-m or fm radiotelephone, and can also use a modulated or interrupted unmodulated carrier for operating radio-controlled devices.

Issuance of Class-B station authorizations was discontinued in March of 1968. Since that time, no frequencies have been made available either for new or modified facilities. However, existing Class-B stations may, upon renewal, continue operation in this class of service until November 1, 1971.

Class-C License

Shortly after the opening of Classes A and B in the 460-470 MHz band, an additional frequency (27.255 MHz) was added for remote-control purposes only. Labeled Class C, it was not authorized for any type of radiotelephone operation whatsoever, but was to be used solely for radio control of such devices as garage doors, model airplanes, boats, etc.

Frequency Allocations—Class C was initially allocated only one frequency (27.255 MHz). However, in 1958, the FCC authorized five additional operating frequencies as follows: 26.995 MHz, 27.045 MHz, 27.095 MHz, 27.145 MHz and 27.195 MHz. The initial frequency of 27.255 MHz is shared by stations in other services, and stations operating on any of these frequencies can expect no protection from interference caused by scientific, medical or industrial equipment operating in the same frequency range.

The most recent Class-C frequency acquisition consists of five channels (72.08 MHz, 72.24 MHz, 72.40 MHz, 72.96 MHz and 75.64 MHz) to be used solely for the remote control of model airplanes, providing such operation does not interfere with the remote control operation of industrial equipment using the same or adjacent frequencies or with television reception on Channels 4 or 5.

Power Limitations—Maximum plate-power input permitted a Class-C station is 5 watts (4-watts output), except on the original frequency of 27.255 MHz where the maximum allowable input is 30 watts (24-watts output).

Emission—Only two types of emission are permitted for Class-C operation—amplitude tone modulation, and on-off unmodulated carrier. No intelligence can be transmitted in this band; it is reserved for remote control uses only.

Class-D License

Class-D Citizens radio has taken the spotlight away from all other classes. In fact, many had never heard of Citizens Band radio until this service was announced.

One of the inherent disadvantages of the Class-A and -B services is that they operate in the uhf region. This presents problems in equipment design and places a considerable limitation on the communicating range. The Class-D service, assigned much lower frequencies, overcomes to a great extent the line-of-sight characteristics in the higher 460-470 MHz band. This not only extends the reliable comunicating range, but also adds to the usefulness of the equipment itself. Furthermore, a large variety of Class-D equipment, capable of meeting almost every need, is now available in either kit or factory-wired form at prices within almost any budget. Most of the Class-D radio equipment is of the transceiver type similar to the units shown in Fig. 1-4. Other examples of this type are contained in Chapter 2.

Frequency Allocations—When the FCC established the Class-D service, it allocated a group of frequencies formerly assigned to the 11-meter amateur band. This assignment gave Class-D stations 22 channels on which to operate exclusively. Since then a 23rd channel (27.225 MHz) has been made available on a shared basis with stations in other services. Fig. 1-5 shows the location of the Class-D

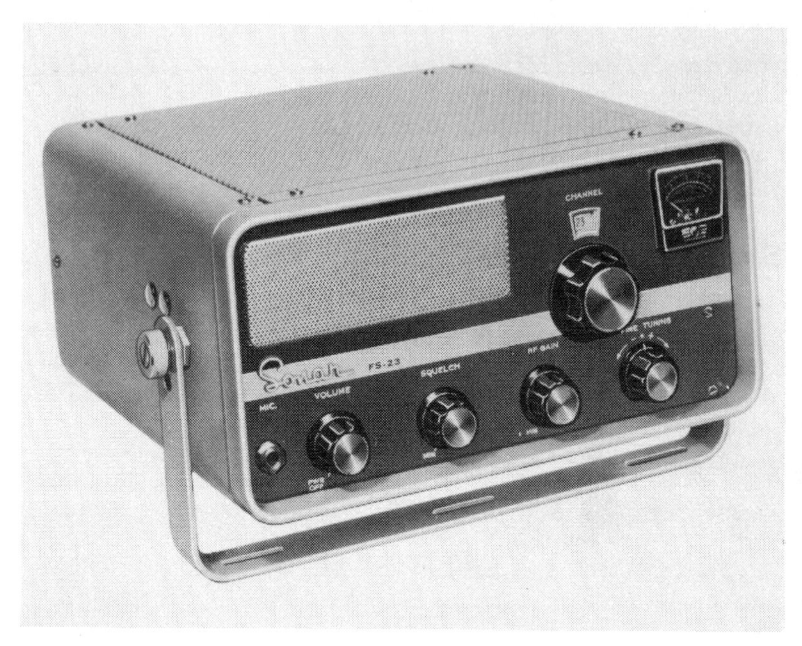

(A) Tube-type transceiver.

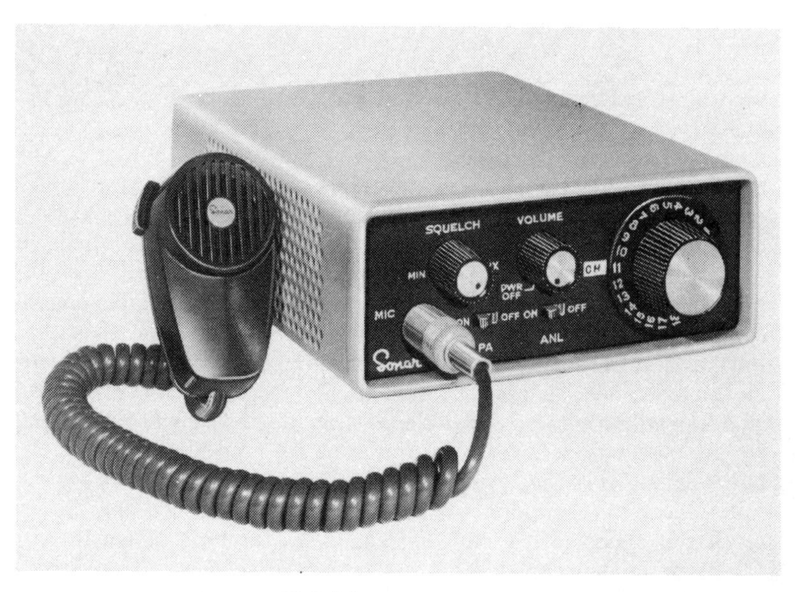

(B) Solid-state transceiver.

Courtesy Sonar Radio Corp.

Fig. 1-4. Typical Class-D equipment that is currently available.

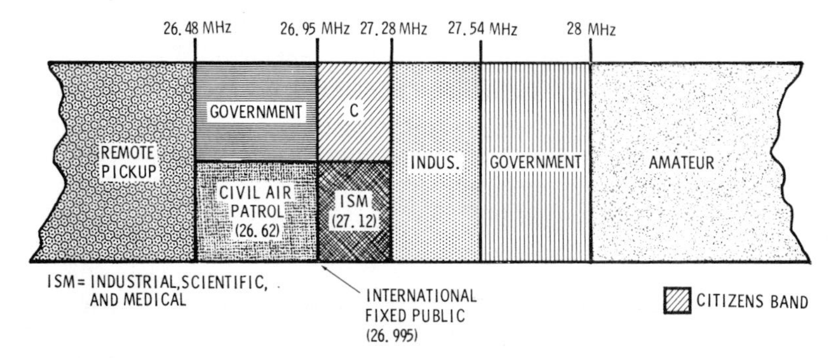

Fig. 1-5. Position of Class-D channels in the radio frequency spectrum.

frequencies in the radio spectrum. Table 1-2 lists all 23 Class-D channels and corresponding operating frequencies. Note that they are 10 kHz apart except where separated by Class-C channels.

Class-D equipment that is not crystal controlled must have an FCC type-acceptance number. (Some transceivers are type-accepted even though they are crystal controlled.) As with Class C, there is no guarantee against interference originated by industrial, medical, and scientific devices operating in the 26.96 to 27.28 MHz band.

Although a total of 23 channels is available to Class-D stations, there are certain restrictions connected with their use. For example, Channels 1 through 8 and 10 through 23 are to be used for communications between units of the same station. In order to communicate with other stations (those under the control of another licensee), such communications must be conducted on Channels 10 through 15 and Channel 23.

The only remaining channel is 9. This channel is reserved solely for emergency communications involving the immediate safety of individuals and the protection of property, or for communications

Table 1-2. Class-D frequencies

Channel	Freq. (MHz)	Channel	Freq. (MHz)	Channel	Freq. (MHz)
1	26.965	9*	27.065	17	27.165
2	26.975	10	27.075	18	27.175
3	26.985	11	27.085	19	27.185
4	27.005	12	27.105	20	27.205
5	27.015	13	27.115	21	27.215
6	27.025	14	27.125	22	27.225
7	27.035	15	27.135	23**	27.255
8	27.055	16	27.155		

*Special restrictions.
**Shared with stations in other services.

necessary to render assistance to a motorist. Proper usage of the Class-D channels is discussed further in Chapter 8.

Power Limitations—Class-D equipment is allowed a maximum dc plate power input of 5 watts and maximum rf power output of 4 watts. Although seemingly limited, this meager input can provide communication over surprising distances. (Power is not the only factor that determines communicating range, as will be discussed later.)

Emission—The only emission permitted for Class-D operation is amplitude-modulated (a-m) radiotelephone. Frequency modulation (fm) is *not* permitted for this reason: a-m transmitters in this service are permitted a bandwidth of only 8 kHz; an fm transmitter would require a frequency spread much greater than this figure. Obviously, the excessive space taken up by fm would cut down the number of channels available within the allocated bandwidth.

2

Citizens Band Equipment

An almost unlimited variety of Citizens band equipment is available today. There are sets for every conceivable type of operation, and many of them can be purchased either in kit form or as factory-wired and tested assemblies. Some ingenious individuals may even consider building their own. Regardless of who designs and constructs the equipment, it must conform with certain FCC requirements before it can be put on the air.

TECHNICAL REQUIREMENTS

Frequency Tolerance

All CB transmitters must operate within a specified frequency tolerance. With Class-A and -B equipment, the frequency tolerance depends not only on the class of station but also on the type of operation (mobile, fixed, or base) and the plate input power. For example, the frequency tolerance of a Class-A fixed or base station is .00025 percent, but mobile equipment in the same class is allowed a tolerance of 0.0005 percent. Class-B equipment is authorized for mobile operation only, and must conform to a frequency tolerance of 0.5 percent if its plate input power is 3 watts or less, and 0.3 percent if over 3 watts. Class-C and -D units must operate with a frequency tolerance of .005 percent whether maximum plate input power is used or not.

Modulation

The rf carrier of an a-m transmitter cannot be modulated over 100 percent. Exceeding this amount not only would distort the signal, but would also introduce frequencies which may extend outside the assigned channels. This, of course, is not permitted. Most CB transmitters therefore include some form of self-limiting to prevent overmodulation. The frequency deviation of all fm transmitters used for CB operation must not exceed ±5 kHz after November 1, 1971.

FCC Type Approval

Type approval signifies that the equipment adheres to FCC technical requirements. Most Citizens band equipment is crystal-controlled to ensure frequency stability. However, equipment that is not crystal-controlled may also be used. Such units must have an FCC type-approval number stamped on an attached nameplate. Manufacturers desiring type-approval for CB equipment must first submit a written request to the Secretary of the FCC. The request will usually not be considered unless at least 100 units are scheduled to be manufactured.

Assuming the request is approved, a working model is then submitted to the FCC laboratory at Laurel, Maryland. Here it undergoes extensive testing to determine its performance capabilities under various conditions, such as prolonged exposure to temperatures of anywhere from 0° to 125° F., and humidities ranging from 20 to 95 percent. Other tests include operation under the voltage variations encountered during normal usage, and also the effect on operation when the position of the equipment and surrounding objects is changed. Occasionally, these tests are performed by a cooperating governmental department rather than by the FCC itself.

Before equipment is approved or rejected, the test results are forwarded to the Commission, which, in turn, confidentially advises the manufacturer of the decision. Naturally, all other production models must duplicate the originals as closely as possible, and the design or construction must not be changed without approval from the Commission. For his own benefit, a manufacturer may request type-acceptance for crystal-controlled equipment.

SPECIFICATIONS

The best way to determine what a particular Citizens band unit has to offer is to examine carefully its specification sheet, which shows at a glance what would ordinarily take hours to determine by an analysis of the equipment itself. However, unless you are able to interpret the meaning of these specifications, they will be of little value to you. If you are not able to check the operation of the equipment before you buy it—and this is usually the case—you must rely on the specifications alone. Chart 2-1 shows a typical specification sheet for a Citizens band transceiver.

Chart 2-1. Typical Equipment Specification Sheet

SPECIFICATIONS

General Information

Power Supply:	12 VDC (115 VAC with accessory adaptor)
Channels:	All 23 channels
Frequency Range:	26.965 to 27.255 MHz
Frequency Control:	Precision quartz crystals
Frequency Tolerance:	0.003%
Operating Current:	Transmit 1.2 amps, full modulation
Size (including mounting bracket):	2½ inches high, 6½ inches wide, 9 inches deep
Weight:	4.5 lbs.

Receiver—Dual Conversion Superheterodyne

Sensitivity:	Less than 1 uV for 10dB S+N/N at 0.5W output, all channels
Selectivity:	15 dB down at 10 kHz, and 60 dB down at 30 kHz
Spurious Response:	70 dB down or greater
Image Rejection:	Better than 45 dB down
Noise Limiter:	Series gate noise limiter
Squelch:	Adjustable carrier-operated (Threshold less than 1 uV. Opens at 250 uV)
Audio Output:	2.5 watts working into 4-6 ohms
Frequency Response:	±3dB, 300-2500 Hz
Distortion:	Less than 5% with 100 uV signal and 0.5W output
Speaker Impedance:	4-6 ohms

Transmitter

Power Input:	5 watts maximum
Power Output:	3 watts, ±0.5, watts nominal
Modulation:	Type A3, A-M. 100% capability (self-limiting)
Frequency Response:	±3dB, 300-2500 Hz
Output Impedance:	50 ohms nominal, unbalanced
Microphone:	High-impedance crystal type with push-to-talk button.

Controls

Controls:	On-off, volume, squelch, channel selector, noise limiter

Many of the entries are self-explanatory. Under "General Information" you will usually find such things as the power supply requirements, number of channels, current drain, physical dimensions, and weight. The entry for operating current often states the amount of current drain during the transmitting mode since this mode reflects the maximum-current condition. The physical dimensions of the equipment are important when space is at a premium, but the weight of the equipment is seldom restrictive.

The next section of the specification sheet concerns the receiver, which our example describes as a dual-conversion superheterodyne covering all 23 Class-D channels. Next is sensitivity, or the ability to receive weak signals and amplify them to a sufficient level. Often this is stated as so many microvolts for a given signal-to-noise ratio, usually 10 dB. This particular receiver is capable of amplifying to a usable level an input signal of less than 1 microvolt (one millionth of a volt.).

The selectivity of a receiver is its ability to reproduce a desired signal while rejecting all others. This characteristic is described in terms of so many dB down at a given frequency, and can be more easily understood by examining the selectivity curves in Fig. 2-1. The peak (maximum response) of the solid-line curve occurs at the de-

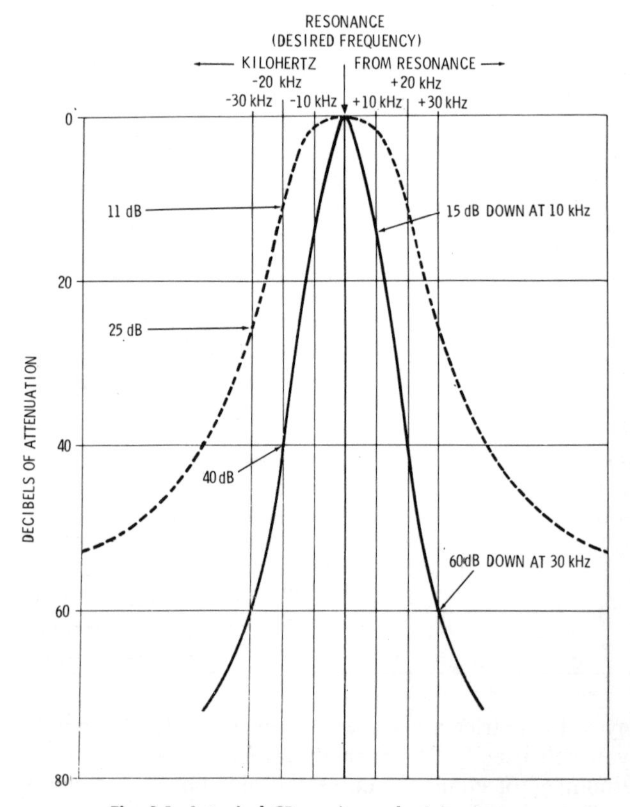

Fig. 2-1. A typical CB receiver selectivity curve.

sired frequency. Notice the steepness of the slope on either side of this maximum point. If another station were located within 30 kHz of this peak, its signal would be attenuated by 60 dB and thus offer negligible interference with the desired signal. The dotted line illustrates a poor selectivity curve. Here, a signal from a station located 10 kHz from either side of the peak would be given practically the same amplification as the desired signal, resulting in two stations being heard at the same time. At 20 kHz from the peak, the adjacent signal would be down approximately 11 dB (compared with 40 dB for the solid curve), and only 25 dB at 30 kHz—still not enough to prevent some interference. The latter response curve is somewhat exaggerated to show how a sharper response curve provides better selectivity—an important consideration in Class-D operation, where the channels are only 10 kHz apart.

Another factor pertinent only to superhet circuits is the amount of image-frequency rejection. Briefly, during frequency conversion, an rf signal from the local oscillator is mixed with the incoming station signal to produce a third (i-f) frequency—actually two frequencies, one above and one below the oscillator frequency. The higher one is the undesired image frequency, which must be attenuated sufficiently to prevent interference. From the specification sheet in Table 2-1, note that the image frequency is attenuated better than 45 dB—enough to make interference negligible.

Also generally listed in the receiver specifications are the type of squelch circuit and its sensitivity, and the type of noise-limiter circuit employed. These circuits are covered in Chapter 3.

The audio-output power of the receiver in our example is 2.5 watts when working into a 4-to-6-ohm load. Maximum power will be transferred from the output of the receiver to the speaker when the impedances of the two are the same. Therefore, using a speaker with an improper impedance would result in loss of power due to the mismatch. The figure in the specification sheet is the maximum power delivered when the audio-output amplifier of the receiver is terminated with the proper load (the 4- to 6-ohm impedance of the speaker).

The transmitter specifications usually include the amount of plate power input to the final stage—in our example, 5 watts (the maximum allowed by the FCC for a Class-D transmitter). You need not be too concerned about the frequency stability of commercial equipment, since it is made to conform with FCC requirements before being sold. (Of course, it will be your responsibility to see that it continues to operate within this tolerance.) Another entry often included is the modulation capabilities. It is desirable for the modulation to approach 100 percent, since to some extent it determines the communicating range. In some instances the type of microphone

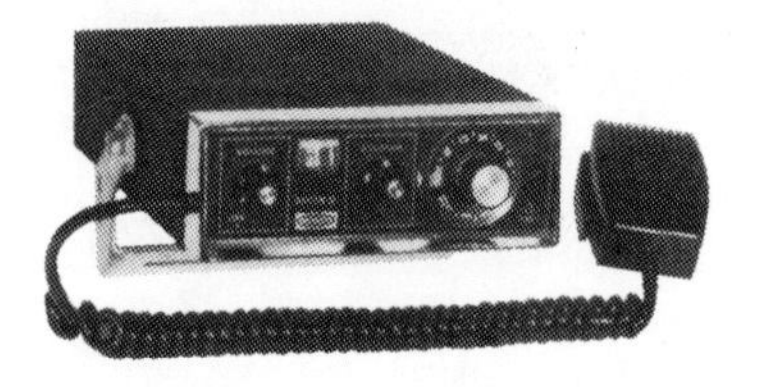

Raytheon Raycom III

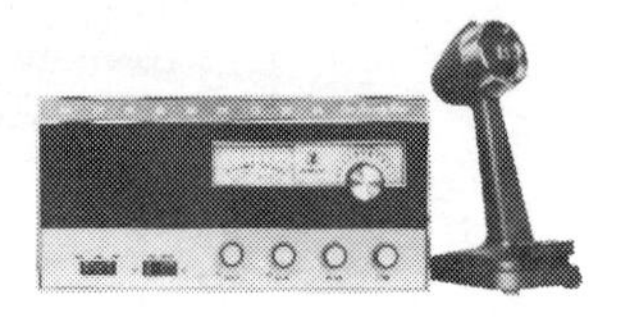

Johnson Messenger 124

Pearce-Simpson Tiger 23

Dynascan Cobra 98

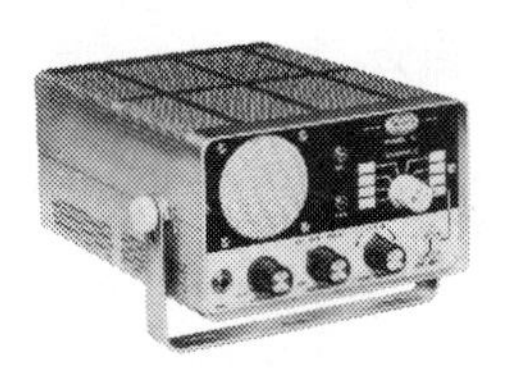

Sonar Model "H"

Pearce-Simpson Bearcat 23

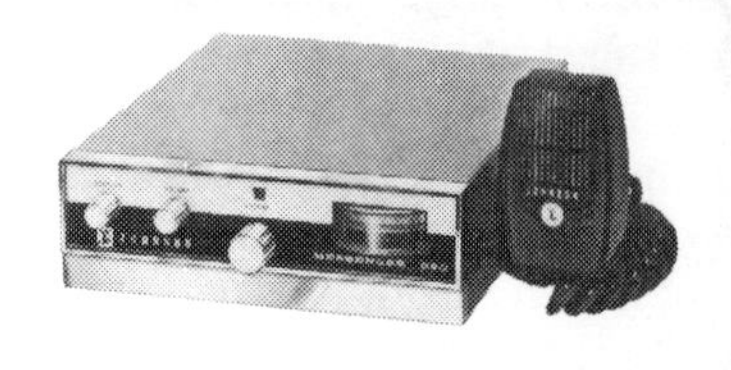

Johnson Messenger 320

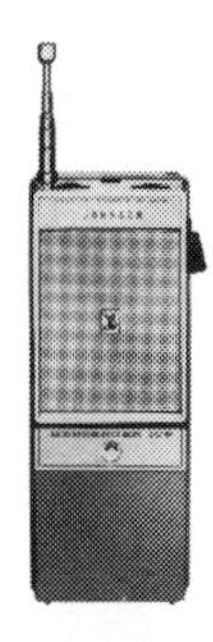

Johnson Messenger 109

General Radiotelephone Model VS-12

Fig. 2-2. Examples of current

Lafayette Model HB-525D

Pearce-Simpson Guardian 23

Dynascan Cobra Cam 88

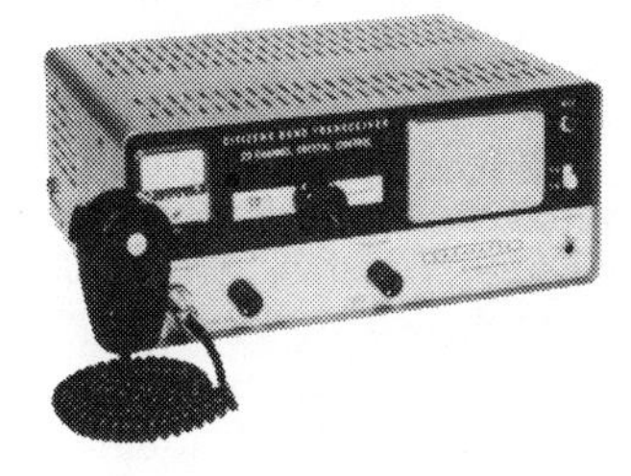

Lafayette Comstat 23 Mark VI

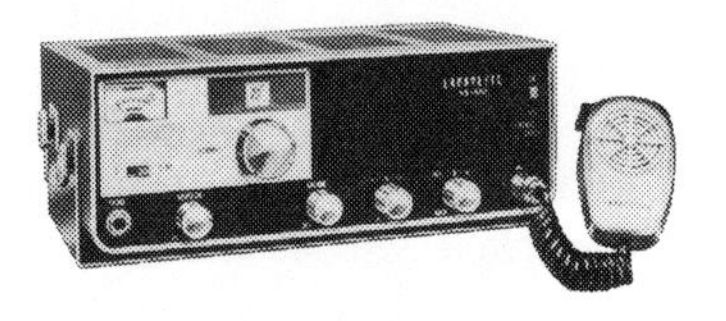

Lafayette Model HB-600

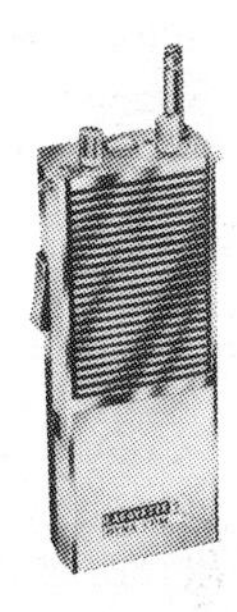

Lafayette Dyna Com 2A

Browning Golden Eagle SSB-15

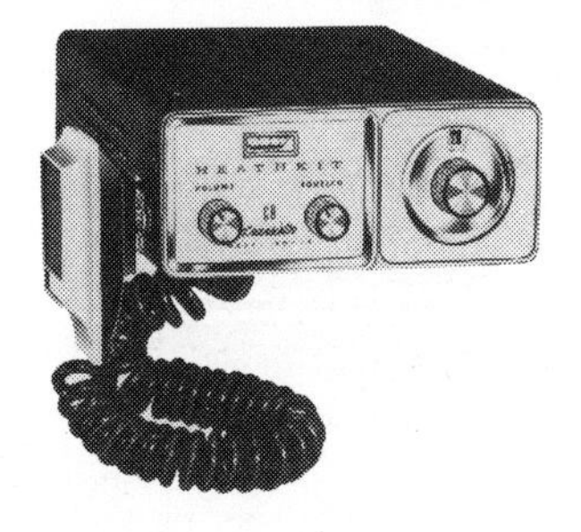

Heath Model GW-14

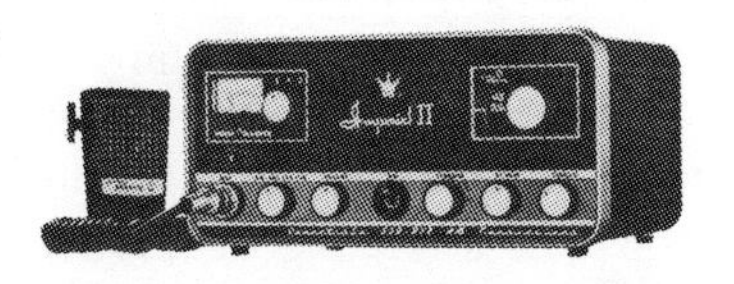

Regency Imperial II

Class-D transceivers.

employed will be specified, as well as the various operating controls (the control functions are fully explained in Chapter 8).

FACTORY-BUILT EQUIPMENT

Before deciding what type of equipment to purchase, it is advisable to examine the offerings of a number of manufacturers. Due to the number of Citizens band equipment manufacturers, it is virtually impossible to describe all of their instruments in this book. However, those which have been included are representative samples of what is available at the present time.

Either fixed-tuned or tunable units, or combinations thereof, are available. Furthermore, most transceivers are provided with a choice of power supplies. Some are designed strictly for 115-volt ac operation; others can be operated from either an ac or dc source. Fig. 2-2 shows a number of typical Class-D Citizens band transceivers that are currently available.

Single-Channel Fixed-Tuned

In single-channel, fixed-tuned, Class-D equipment, both the transmitter and receiver are designed to operate on a specific channel. The transmitter (and usually the receiver) are crystal-controlled to ensure frequency stability—at least within the specified tolerance.

Although a transceiver of this type can operate on only one channel at a time, it can easily be made to operate on any of the others in this class by simply inserting the appropriate crystals.

Class-A stations require prior approval from the FCC before changing operating frequencies; Class-D stations, however, can operate on any of the 23 channels (within limitations) without such approval.

Few single-channel base and mobile transceivers are being produced at this time. Most now are of the multichannel variety. The majority of the small hand-held units are the single-channel type, however.

Multichannel Fixed-Tuned

Perhaps the equipment is to be used in an application where several operating channels are needed, but with some means of instantly switching the proper crystals into the transmitter and receiver circuits. The multichannel, fixed-tuned transceiver is the answer. It can be either a unit in which a single control simultaneously selects transmitter and receiver crystals for the same channel, or one employing separate transmitter and receiver channel selection.

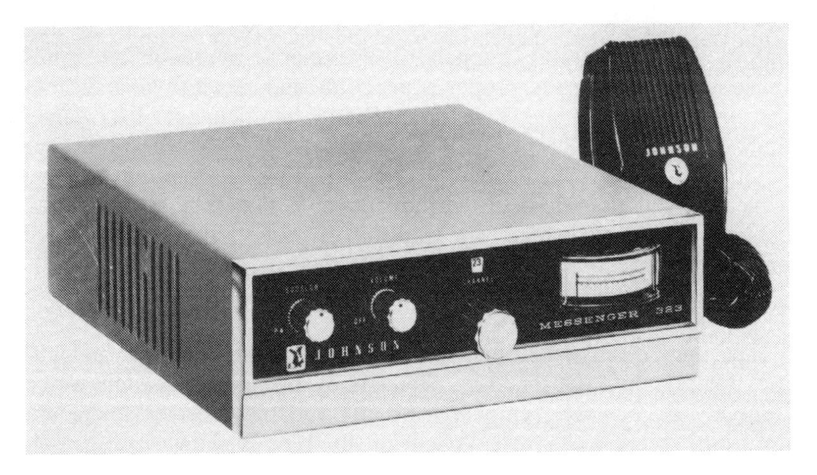

Courtesy E. F. Johnson Co.

Fig. 2-3. The "Messenger 323" is a mobile transceiver providing switch selection of all 23 channels.

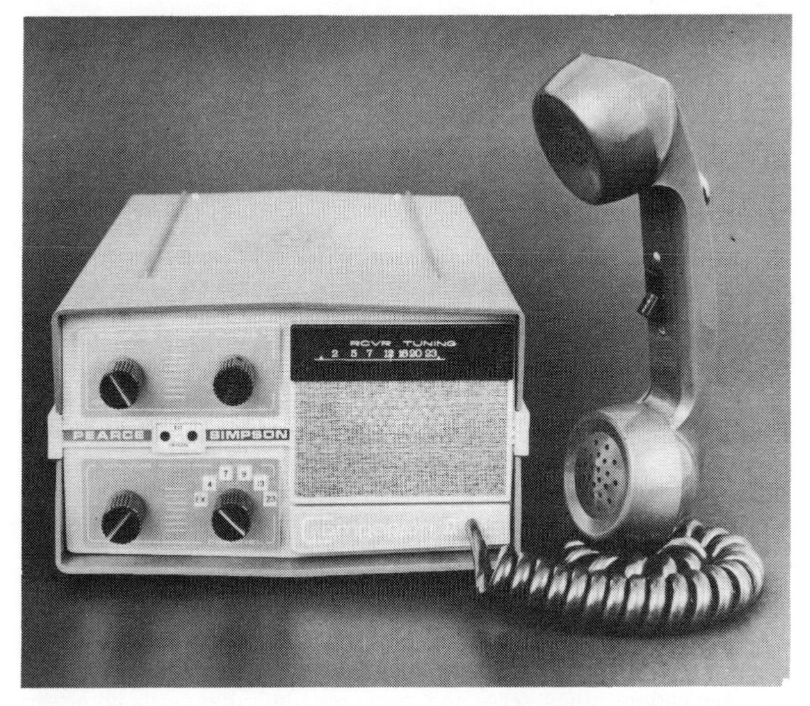

Courtesy Pearce-Simpson, Inc.

Fig. 2-4. The "Companion II" incorporates a six-channel crystal-controlled transmitter and a continuously tunable receiver.

With single-control units, a selector on the front panel enables any one of several Class-D channels to be put into operation by simultaneously switching the proper transmitter and receiver crystals into the circuit. Fig. 2-3 shows an example of a multichannel fixed-tuned unit capable of receiving all 23 channels. Transceivers that are equipped with separate controls for selecting transmitter and receiver crystals provide a further advantage in that they permit cross-channel operation, where you transmit on one channel and receive on another. (See Chapter 8.)

Tunable

The units discussed so far have employed a fixed channel or facilities for selecting one or more channels by means of a switch. Some transceivers, however, employ a strictly tunable receiver; there are no preset receiver channels. A unit of this type is shown in Fig. 2-4. This dual-control feature is a requirement for engaging in cross-channel operation.

Combination Fixed-Tuned and Tunable

A popular variation of the two previously discussed designs has resulted in the combined fixed-tuned/tunable arrangement. Transceivers utilizing this method are equipped with a receiver that is crystal-controlled on a given number of channels but also continuously tunable. Selection between the two operating modes is normally accomplished with a single control labeled XTAL/TUNE.

Solid-State Equipment

Transistors, integrated circuits, and other solid-state devices have made possible many changes in electronic equipment. Two distinct advantages of the solid-state concept over vacuum tubes are a great reduction in size and lower power consumption. When CB radio was first introduced there was little, if any, solid-state equipment available. Now, however, more than half of the equipment, ranging from the full-powered base and mobile stations to the small hand-carried "walkie-talkies," employ solid-state circuitry. Some units use solid-state exclusively (even for switching purposes) and others in only part of the circuits. Equipment using both tubes and solid-state devices is commonly referred to as "hybrid." Fig. 2-5 illustrates a typical "all solid-state" Class-D transceiver.

Hand-held transceivers are almost exclusively solid state, and the majority operate at relatively low power, usually 1 watt or less. Since the primary function of these smaller units is to provide short-range communications, they serve the purpose admirably. Depending on surrounding terrain and other factors, it is not uncommon for such units to provide reliable communications up to two miles or more.

Courtesy Dynascan Corp.

Fig. 2-5. The "Cobra 24" is a typical compact solid-state transceiver with a full 5-watts input on all 23 channels.

Fig. 2-6. The "Dyna-Com 12" is a hand-held transceiver featuring 12-channel operation and a full 5 watts of input power.

Courtesy Lafayette Radio Electronics Corp.

Furthermore, no FCC license is required when the transmitter power input is 100 milliwatts (one-tenth watt) or less. This class of equipment, however, must be used for communications with similar unlicensed equipment only.

Although most hand-held transceivers operate with relatively low transmitter power and provide only one or two operating channels, the designs are beginning to change. For example, Fig. 2-6 shows a 12-channel hand-held transceiver that operates at the maximum 5 watt transmitter power input. With accessories, this transceiver will also operate from 12 Vdc in mobile use or from 110 Vac when used as a base station. The overall size of this unit is 3¼″ wide, 10½″ high, and 2$\frac{1}{16}$″ deep.

Although the majority of Class-D Citizens radio equipment is of the transceiver variety (transmitter and receiver combined with some circuits common to both), separate units are also available as illustrated by the example in Fig. 2-7. Here, the transmitter and receiver are completely separate units. Several manufacturers are currently producing equipment in this form. Thus, another choice is provided for prospective buyers.

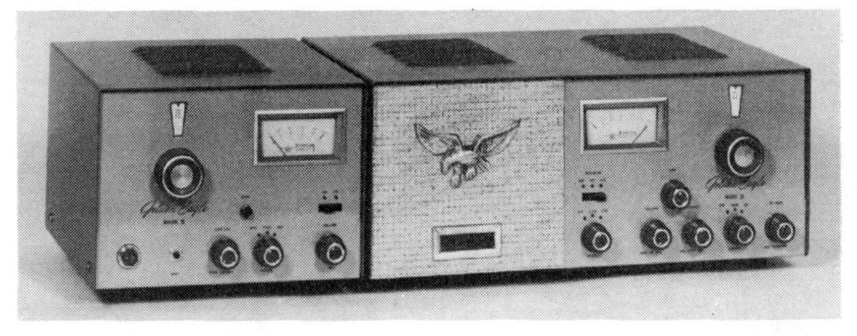

Courtesy Browning Laboratories, Inc.

Fig. 2-7. The "Golden Eagle Mark II" is an example of a Class-D transceiver utilizing a separate transmitter and a separate receiver.

KITS

CB transceivers are also available in kit form. Critical circuits (such as the oscillator)—which might cause illegal operation if not assembled and adjusted correctly—are wired, pretuned, and sealed at the factory. This procedure is necessary to conform with FCC regulations and to make it possible for a person without a commercial radiotelephone license to assemble the unit. Some kits are assembled from individual components, while others are made up of prewired subassemblies that are put together by the builder. In practically all kits, the critical circuits have been pretuned and sealed at the factory to insure legal operation.

SPECIAL AND ACCESSORY EQUIPMENT

In addition to the basic Citizens band equipment (transmitter and receiver), there are a number of accessories. Some of these are required for station operation—others are merely refinements to improve performance, add versatility, or in some other way enhance the equipment.

RF Converters

Basically, the rf converter is a tuner (usually containing its own power supply) designed to cover a specific band of frequencies. Used with a standard a-m broadcast receiver (either home or automobile type), it selects the desired signal and converts its frequency into one which can be tuned in on the broadcast receiver. Some CB equipment is available with a separate transmitter and receiver. Hence, should you have a transmitter but no receiver, or if you simply desire an additional receiver (to monitor CB transmissions in your car, for example), the rf converter offers the ideal solution.

Transverters

The transverter, as its name implies, is a combination transmitter and converter. The transmitter section must be crystal-controlled if it is to be used for Class-D operation, and must conform in all other ways to the FCC regulations governing equipment design. The converter section is operated in conjunction with a standard broadcast receiver as described previously.

Coded Calling Systems

The CB bands are becoming very crowded. Listening to all the other stations to make sure that you receive your calls can be very tiring. Consequently, methods have been devised to prevent all but the desired signals from being received.

One type of coding system, called tone squelch, is provided on some CB equipment to reject all transmissions other than those within a controlled network. Each station within the network transmits a coding tone in addition to the voice signal. Some systems transmit a tone (made inaudible by filters) during the entire message; others transmit only a brief tone when the mike button is first depressed. Each receiver is equipped with a decoder that responds only to the tone transmitted from stations within the network.

Some transceivers contain a built-in tone squelch system; with others it can be added as an accessory. Figs. 2-8 and 2-9 illustrate

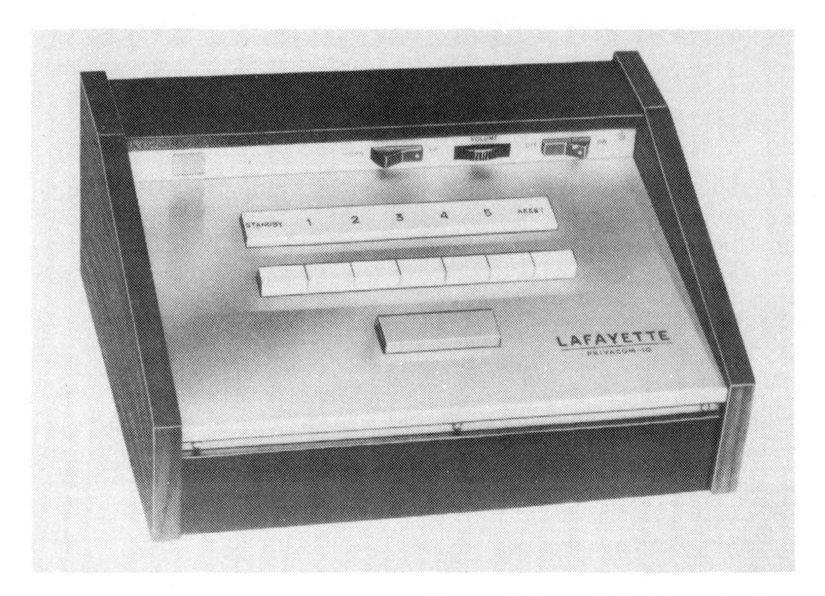

Courtesy Lafayette Radio Electronics Corp.

Fig. 2-8. The "Priva-Com 10" is a tone calling system that permit selective calling of one to ten stations simultaneously.

Fig. 2-9. The "Tone Alert" selective calling system.

Courtesy E. F. Johnson Co.

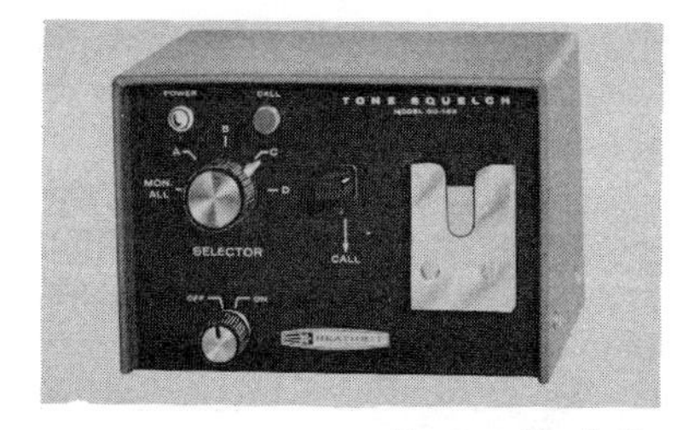

Fig. 2-10. Heathkit tone squelch kit Model GD-162.

Courtesy Heath Co.

two types of add-on units. Another tone squelch device is illustrated in Fig. 2-10. This system employs a resonant-reed relay that provides four different tone signals plus a MONITOR ALL position. An indicator lamp on the front panel lights and an audible tone is heard from the transceiver when a call is coming in. In addition, provisions for activating signaling devices, such as a horn, buzzer, bell, etc., are provided. One tone squelch unit is required for each transceiver. The selector switch on the front panel permits any one of four tones to actuate the transceiver. In the MONITOR ALL position, all tones are monitored and any of them will permit messages to be received.

Speech Limiters and Audio Compressors

The primary function of speech limiters and audio compressors is to permit as high a level of modulation as possible while at the same time preventing overmodulation. The modulation percentage is at its highest on voice peaks, for example as you emphasize a word or syllable. At the same time, however, the average audio level may be well below the 100 percent value. Thus, by amplifying the average audio level and at the same time either clipping or compressing the audio peaks to prevent overmodulation, more "talk power" can be derived from a CB transceiver. Increasing the amount of audio (as

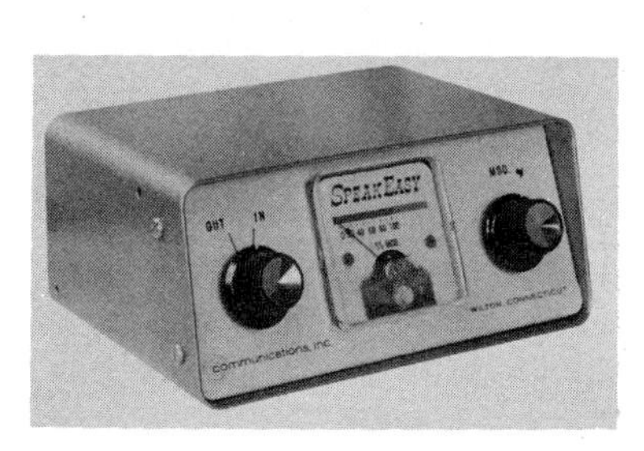

Courtesy Instruments and Communications, Inc.

Fig. 2-11 The "SpeakEasy" audio compressor amplifier.

Courtesy GC Electronics, Div. Hydrometals, Inc.

Fig. 2-12. Globe Model FCL-1 "Speech Booster".

long as it is below 100 percent) definitely improves communications. Therefore, it is desirable that the modulation percentage approach 100 as nearly as possible.

Fig. 2-11 shows one type of speech compressor amplifier. This device compresses the peaks and at the same time "rides gain" to prevent undermodulation. In addition, it incorporates a meter that provides a visual indication of modulation percentage. The *Speech Booster* (Fig. 2-12) increases the overall amplitude of the signal and then *clips* rather than compresses the tops and bottoms of peaks that exceed a preset level. It also attenuates frequencies above 2500 hertz and below 300 hertz.

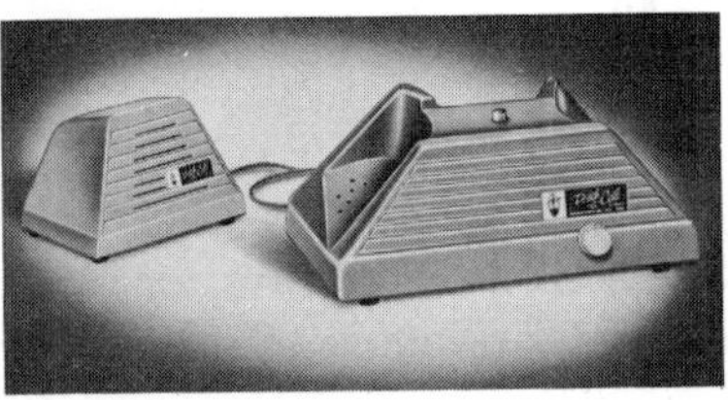

Courtesy Business Radio Inc.

Fig. 2-13. The "Patch-A-Call" Model 301 is designed to permit telephone-to-mobile communications.

Phone Patches

Phone patches are also available for CB radio stations. Basically a phone patch (Fig. 2-13) is a device that links landline telephone to a radio system. When such communications are desirable, the base station operator places the call and then lays the telephone handset into the receptacle provided, automatically turning the unit on. A voice-operated relay automatically keys the base station transmitter as the party on the telephone talks. When he pauses, the station automatically switches to the receiver mode and replies from the mobile unit can be received. An external speaker permits the base station operator to monitor both sides of the conversation.

MICROPHONES

The microphone is an important part of any communications system. It converts sound variations into the corresponding electrical impulses which are to be conveyed by the transmitter. Microphones come in many shapes and sizes (Fig. 2-14); however, the most important part is the inner element.

Dynamic

The dynamic, or moving-coil, type microphone is internally constructed as shown in Fig. 2-15. A coil connected to a diaphragm is

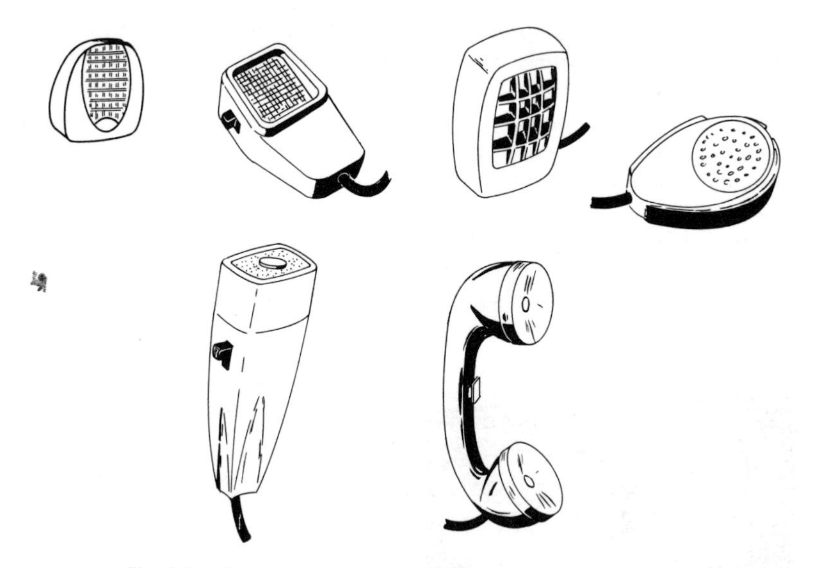

Fig. 2-14. Various types of microphones used in CB equipment.

free to move over a permanent-magnet pole piece. Sound waves entering the microphone cause the diaphragm and coil to move. The movement of the coil through the magnetic field produces the signal voltage.

In some CB equipment, the speaker also serves as the microphone. The dynamic speaker (Fig. 2-16), like the microphone, has a moving coil mounted over a magnetic pole piece and operates on the same principle. During reception, however, the electrical signals pass through the coil and set up a varying magnetic field around it. The interaction between this field and the one generated by the permanent magnet causes the speaker cone to move, producing the sound waves.

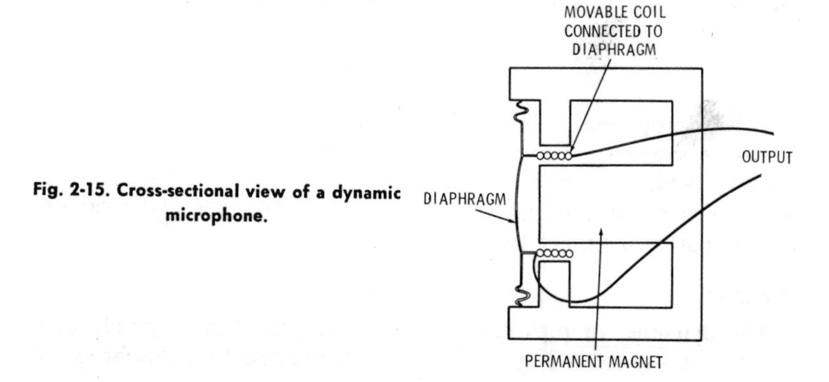

Fig. 2-15. Cross-sectional view of a dynamic microphone.

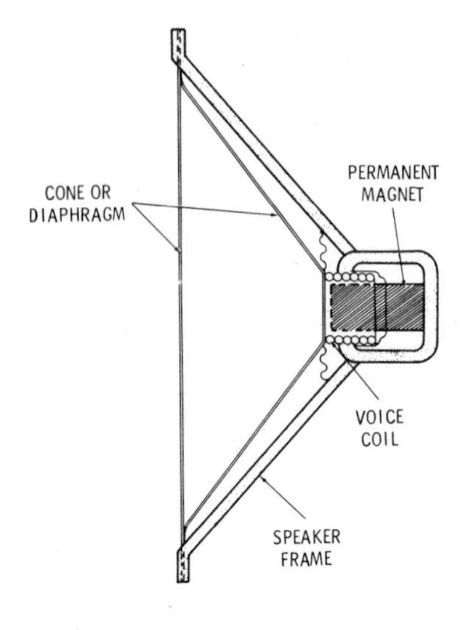

Fig. 2-16. Cross-sectional view of a pm dynamic speaker.

Crystal or Ceramic

Some CB microphones employ a crystal or ceramic slab as the voltage-generating element. The theory of operation is basically the same for both types, as shown by the cross-sectional view in Fig. 2-17. An inherent characteristic of the ceramic or crystal slab is its ability to generate a minute voltage when twisted or bent. As the sound waves move the diaphragm back and forth, the slab bends and produces a corresponding voltage. Crystals cannot withstand prolonged temperatures above approximately 120°F. Ceramic elements, however, are not affected to any great extent by excessive heat.

Carbon

The carbon microphone (Fig. 2-18A) employs the variable-resistance method of operation. Instead of generating a voltage, a dc

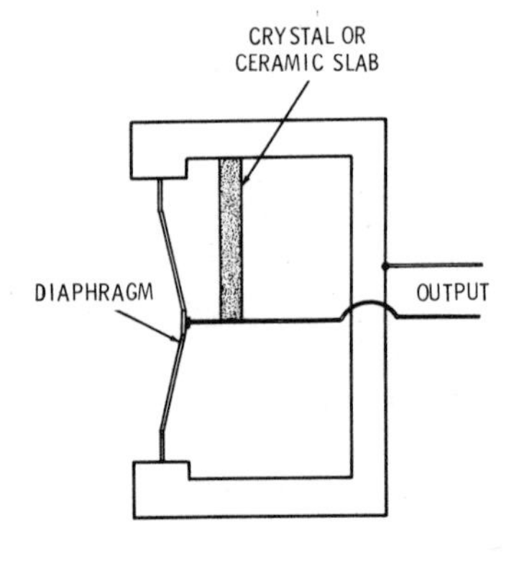

Fig. 2-17. Basic construction of a crystal or ceramic microphone.

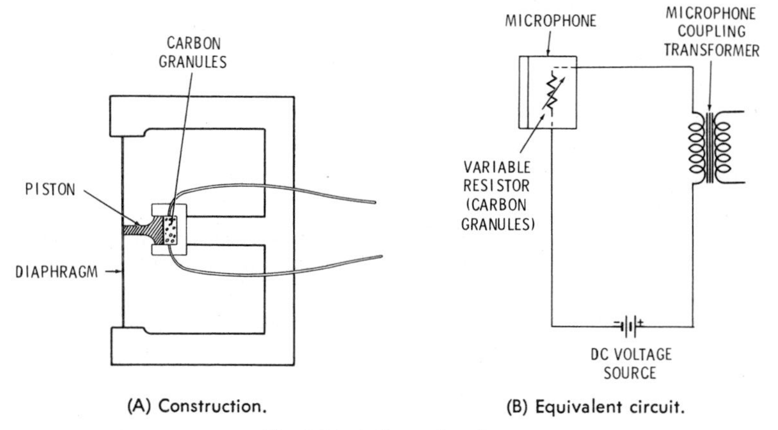

(A) Construction.

(B) Equivalent circuit.

Fig. 2-18. Carbon microphone.

voltage is applied across carbon granules. Any change in the resistance offered by the carbon will affect the direct current accordingly (Fig. 2-18B). Sound waves striking the diaphragm set it and the attached piston into motion. This action varies the pressure applied to the carbon "button," as it is called. This variation in resistance varies the direct current in accordance with the frequency and intensity of the sound waves.

Other Considerations

Not all microphones are as simple as those illustrated here. Some, for example, contain built-in transistor amplifiers to increase audio gain. Thus, when this microphone is used, you are adding another stage of amplification. They may also have controls for varying the tone quality and the amount of output.

The output impedance differs for various microphones. Therefore, it is important that replacement microphones have the same basic impedance characteristics, signal output, etc., as the original mike provided by the manufacturer.

EQUIPMENT SELECTION

In selecting a CB unit, the first consideration is the power source. Make sure that the unit will operate on the type power (6- or 12-volts dc, or 117-volts ac) that you plan to use now and in the future.

Another factor to be considered, when choosing equipment, is the terrain surrounding the area in which it is to be used. A transceiver having less than the maximum allowable plate input power would hardly be satisfactory where maximum distance is desired, yet the same unit might be entirely adequate for shorter distances—particularly over level terrain or water. Hunters, foresters, or anyone requiring short-distance communications away from normal power facilities will find a small transistor portable with self-contained batteries more desirable than a larger transceiver with its cumbersome battery supply. Also, in areas where channel traffic is very light, the receiver need not be too selective—but you should consider the possibility that the channel may some day become quite crowded. A receiver with good selectivity as well as squelch and noise-limiter circuits is a requisite for good reception in the larger metropolitan areas where channel traffic is heavy and the noise level is excessive. If the transceiver is to be used in a car, a mike with a push-to-talk button allowing one-hand operation, is a must.

As you can see, there are many questions to be answered before you can make a satisfactory selection. The pointers offered in this chapter will serve to guide you in selecting the equipment best suited for the job.

3

Receiver and Transmitter Circuitry

The previous chapter covered only the physical aspects of Citizens band equipment. Here we are concerned with actual transmitter and receiver circuits and their principles of operation. Most CB equipment today is of the transceiver variety, in which both the transmitter and receiver are combined on a single chassis and one or more stages are common to both sections. Therefore, it is desirable to first have some understanding of how these sections function as a unit before discussing individual circuits.

In the following discussion, the transceivers are classified according to the type of receiver circuit. Basically, there are two types employed in CB transceivers—superregenerative and the more common superheterodyne. The operation of each will be discussed separately.

SUPERREGENERATIVE TRANSCEIVERS

Very few CB receivers today employ the superregenerative type circuit. However, it was quite popular a few years ago.

The heart of this transceiver is a supperregenerative detector circuit used in the receiver section. In this circuit a controlled amount of rf energy from the plate is fed back, with the proper phase, to the control-grid circuit, causing regeneration and thereby greatly increasing the amplification. This feedback must be properly controlled; otherwise the entire stage will oscillate. A quenching frequency is generated at regular intervals within this circuit to prevent such oscillation. (A separate quenching stage can also be used, but the self-quenching detector seems to be the most popular in CB equipment.)

The superregenerative circuit provides fair sensitivity with a minimum of components. An interesting characteristic of this circuit is its tendency to "hang on to" a signal, thereby acting much like an afc (automatic frequency control) circuit.

SUPERHETERODYNE TRANSCEIVERS

The superheterodyne circuit is more complex than the superregenerative type, and provides the ultimate in sensitivity and selectivity. This circuit, used almost exclusively in CB transceivers, derives its name from the heterodyne method employed for frequency conversion. Basically, this conversion process is accomplished by mixing an rf signal (produced by a local oscillator within the receiver) with the desired incoming station signal. In doing so, a beat frequency different from either of the originals is produced. Known as the intermediate frequency (i-f), it is the difference frequency between those of the oscillator and the incoming signal. (Actually, other frequencies are produced during this process, but here we need consider only the desired intermediate frequency.) Two advantages of the superhet circuit are the greater sensitivity and selectivity that can be achieved when only one frequency (the i-f) is amplified. The frequency of the local-oscillator signal is always different from that of the station signal (either above or below) by the i-f value. Therefore, the frequency of the signal to be amplified is always the same after conversion, regardless of the channel being received.

The i-f signal from the converter is next fed through one or more intermediate-frequency amplifiers to a detector stage. Here the audio signal is separated from the carrier and sent on to the audio amplifier, where it is given the additional boost in power required to operate the speaker.

Single-Conversion Superheterodyne

Fig. 3-1 shows a block diagram of a Class-D transceiver incorporating a crystal-controlled transmitter and a single-conversion tunable superheterodyne receiver. To simplify the analysis, it is desirable to view the stage layout as a block diagram. Switch M1 represents the push-to-talk button. In the receive (R) position, the antenna is connected to the receiver circuits. At the same time, power is supplied only to the superhet tuner and audio-amplifier stages, and the speaker is connected to the circuit, thereby completing the chain. Notice that the receiver is comprised of two sections—the rf tuner and the audio section—split to show how the audio portion is utilized as a speech amplifier and modulator when transmitting. When

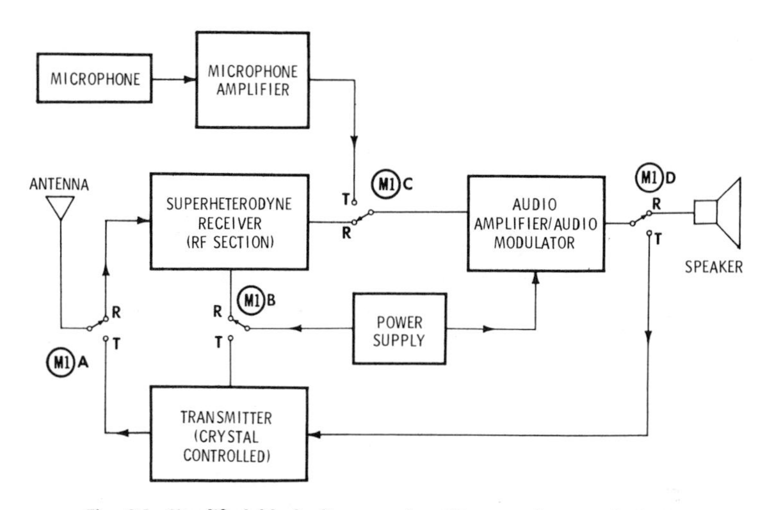

Fig. 3-1. Simplified block diagram of a CB transceiver employing a superheterodyne receiver.

M1 is in the transmit (T) position, the antenna is connected to the transmitter, and the power-supply output is switched from the receiver to the transmitter. The speaker is disabled and the microphone and preamp are connected to the audio system. Inasmuch as the audio section of a transceiver serves as part of both the receiver and the transmitter in most instances, it receives power during either operation. The voice signal follows a path from the microphone to its amplifier, through the audio-amplifier section (which now acts as a speech amplifier and modulator), then on to the final transmitter stage.

Fig. 3-2 shows the schematic for the superheterodyne receiver represented in Fig. 3-1 (the power supply is omitted to simplify the analysis). When switch M1A is in the receive position, the input signal is fed into L1 where it is stepped up to some degree by transformer action before being applied to the grid of rf amplifier V1. The amplified signal is developed across resonant plate load C6-L2, and then coupled via C8 to the grid of mixer tube V2A. Transformer L1 and resonant circuit C6-L2 are tuned broadly enough to accommodate the entire range of Class-D frequencies. The triode section of V2 acts as the local oscillator, which operates in conjunction with a tank circuit comprised of L3, C12, C13, and variable capacitor C14 (the receiver tuning control). The oscillator frequency tracks below the incoming signal by 1750 kHz—the i-f frequency. Hence, the oscillator frequencies range from 25.215 to 25.505 MHz, corresponding to the 1750-kHz difference between the Class-D channel frequencies extending from 26.965 to 27.255 MHz. C9 couples the proper amount of oscillator signal to the grid of V2A. The mixer output consists of the rf input signal, plus the sum and difference frequencies of the oscillator. However, the first i-f transformer (T1), and all those succeeding it, are tuned to the 1750-kHz difference frequency and will pass and amplify only this frequency.

As tuning capacitor C14 is rotated through its range, all Class-D frequencies 1750 kHz above the local-oscillator frequency will be received and converted to the 1750-kHz i-f signal. This signal is amplified by V3A (coupled from the mixer-oscillator stage by i-f transformer T1), and then by V4A (coupled from V3A by T2). A

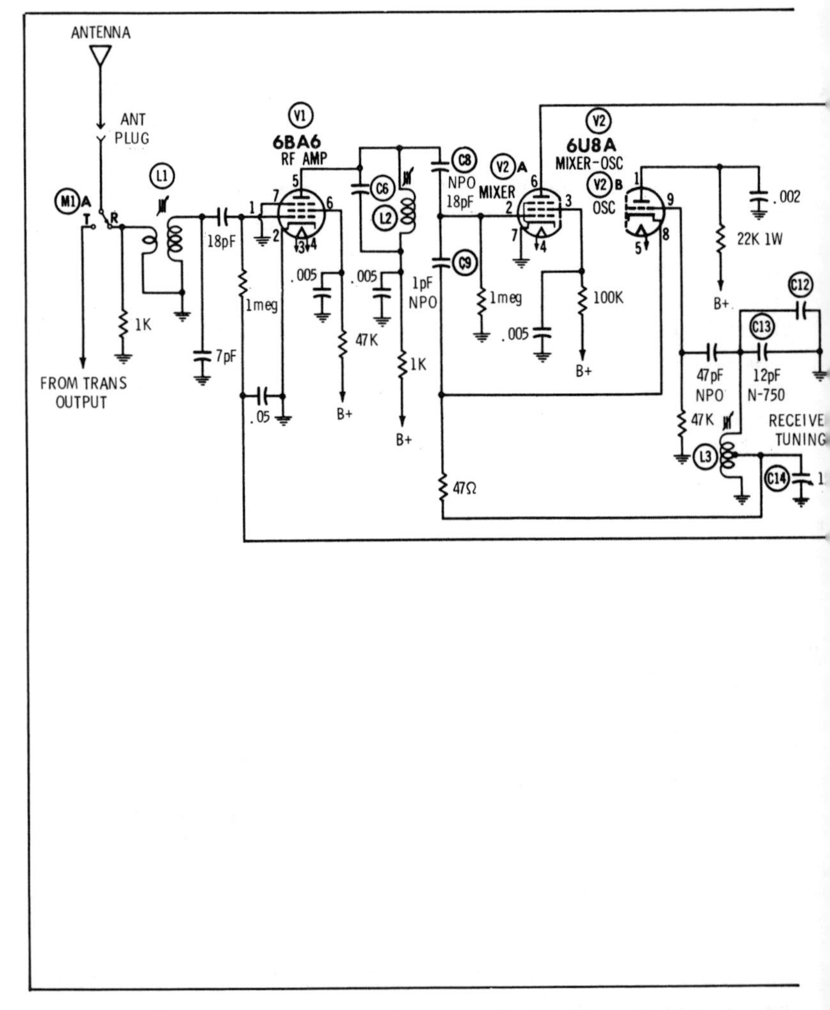

Fig. 3-2. Schematic of the superhet-

third i-f transformer (T3) feeds the signal to the first section of V5A (pins 1 and 6) which serves as a series diode detector. The second diode section of this tube (pins 2 and 3) and its associated components form a series noise-limiter circuit that filters out interference.

The detected audio signal is developed across volume control R36 and applied to first audio amplifier V4B. This stage, in conjunction with V6, operates as a conventional triode voltage amplifier driving a pentode output stage. Notice that audio-output transformer T6 has a tapped primary which enables it to serve as a modulation transformer when transmitting.

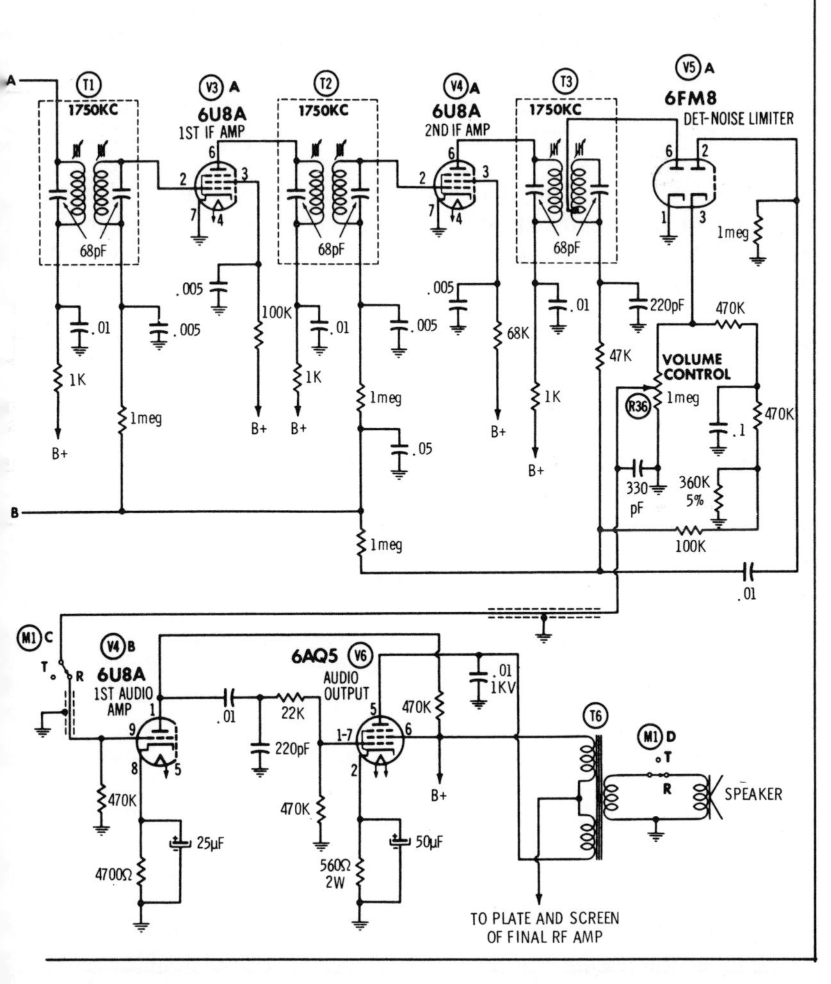

erodyne receiver in Fig. 3-1.

Dual-Conversion Superheterodyne

Another superhet circuit, widely used in Citizens-band equipment, incorporates two frequency-conversion stages. The additional stage provides greater selectivity (rejection of unwanted signals on adjacent frequencies).

Fig. 3-3 shows the stage arrangement in a typical Class-D transceiver employing a dual-conversion superhet receiver. The signal, after entering the antenna, is fed into a conventional rf amplifier stage, where the approximate 27-MHz channel frequency is selected and given initial amplification. It is then injected into the first mixer stage and mixed with the proper oscillator frequency to produce a beat signal at the desired high-i-f value. This signal, in turn, is fed into the second mixer stage where a similar action occurs, this time producing the desired low-i-f frequency. In some receivers, the local oscillator and mixer are combined within a single stage which is termed a converter.

The remainder of the stages operate much like the single-conversion circuits discussed previously: When switch M2 is in the receive (R) position, stages *A* and *B* are part of the receiver circuit. With M1 in the transmit position, these two stages no longer act as part of the receiver, but combine with stage *C* to form the transmitter.

SOLID-STATE TRANSCEIVERS

Transistors offer many advantages over tubes, particularly in mobile and portable CB equipment where current consumption and physical size are important. Some units use transistors, integrated circuits, and other solid-state devices exclusively; others employ both vacuum tubes and transistors, the latter primarily in the power supply and output stages.

A schematic of an all-transistor portable transceiver appears in Fig. 3-4. This circuit employs eight transistors and four semiconductor diodes. No rf amplifier stage is used in this particular application; the received signal at the antenna is connected directly to L1 in the mixer circuit. A signal from crystal-controlled local oscillator X1 is injected into the base-emitter circuit of mixer stage X2 to produce an i-f signal. Two succeeding i-f stages amplify the signal before it is detected by diode M2, which also provides the avc voltage. The audio signal is developed across volume control R18 and passed through audio-amplifier stages X7 and X6 before being coupled to the speaker by T5. Earphone jack J1 provides private listening.

Another interesting feature is that the speaker also doubles as the microphone when switch M5D is in the transmitting (T) position.

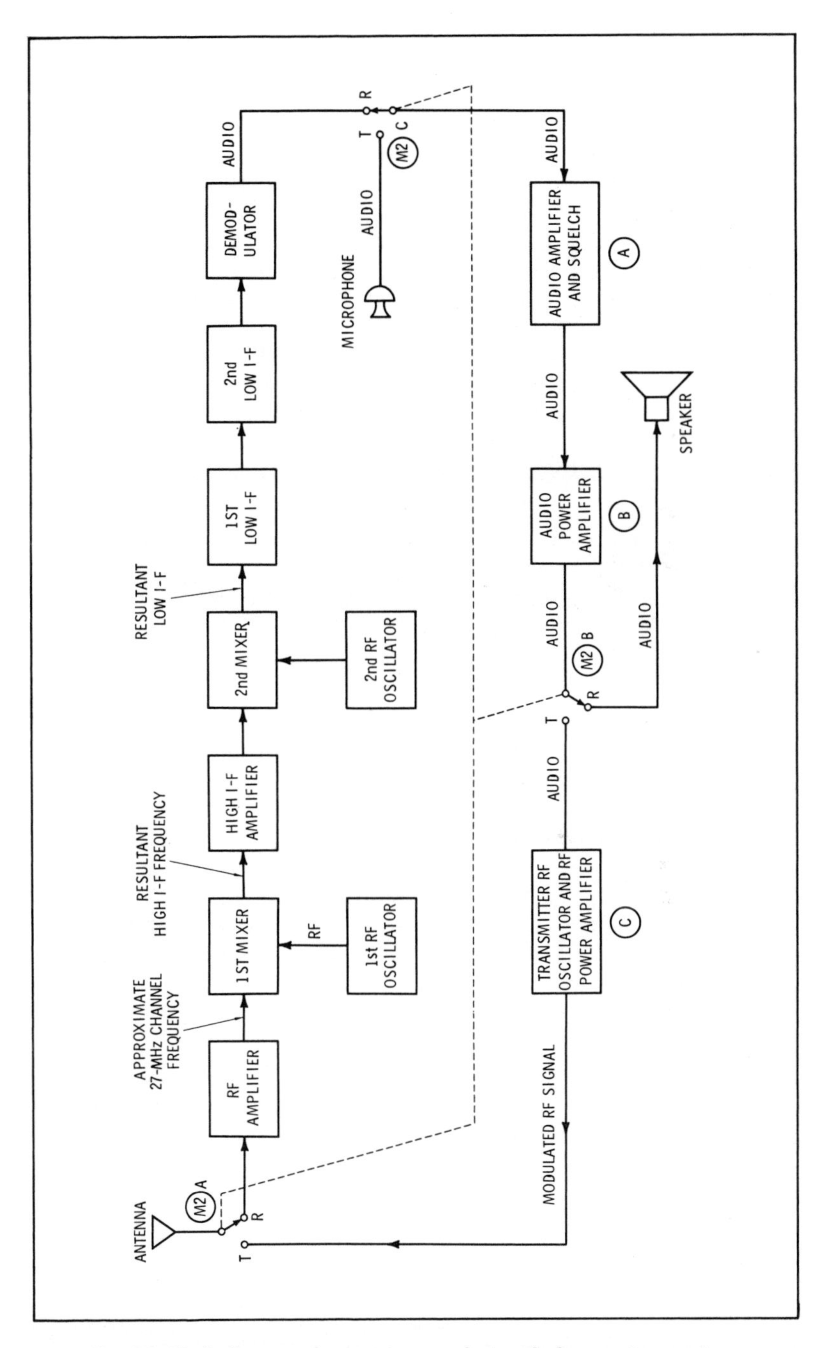

Fig. 3-3. Block diagram of transceiver employing dual-conversion receiver.

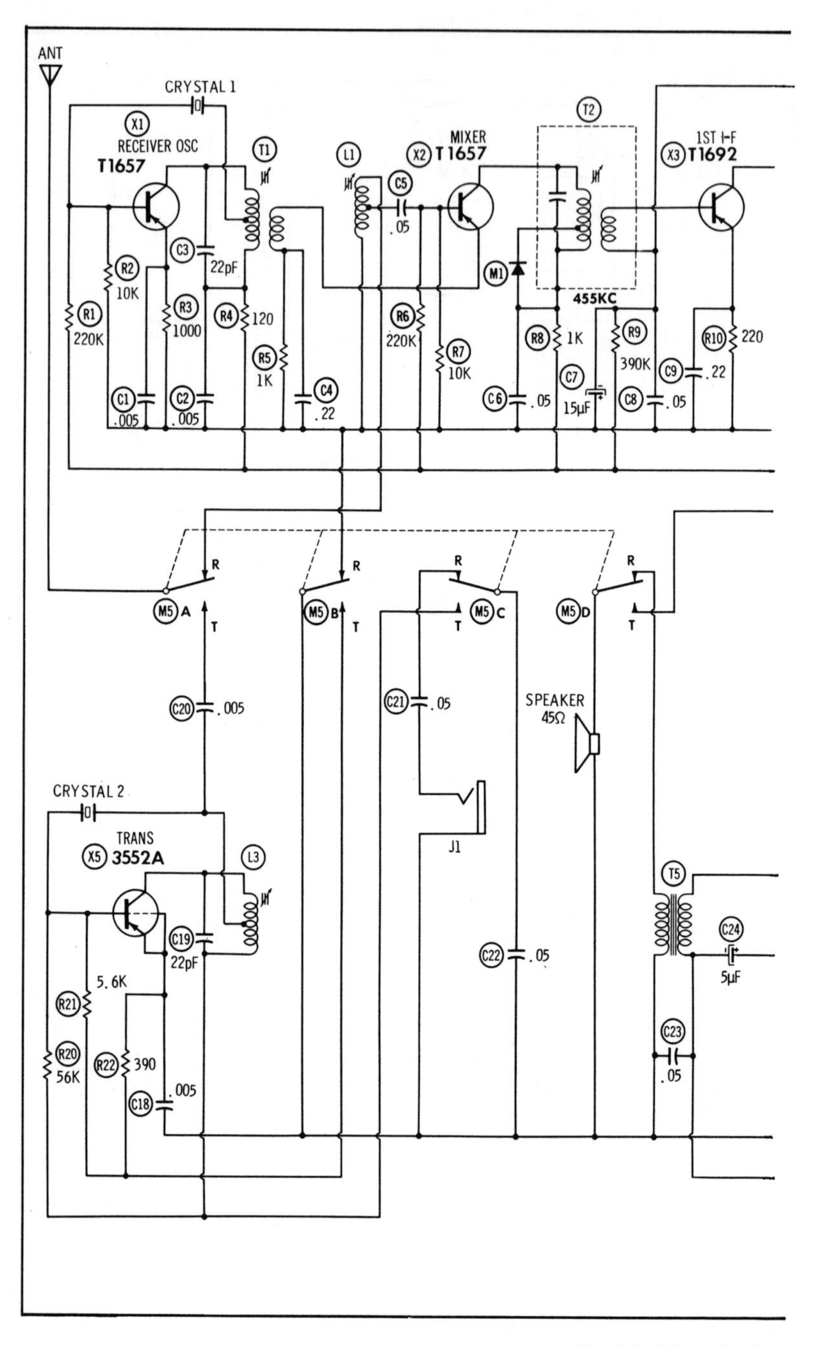

Fig. 3-4. Schematic of an

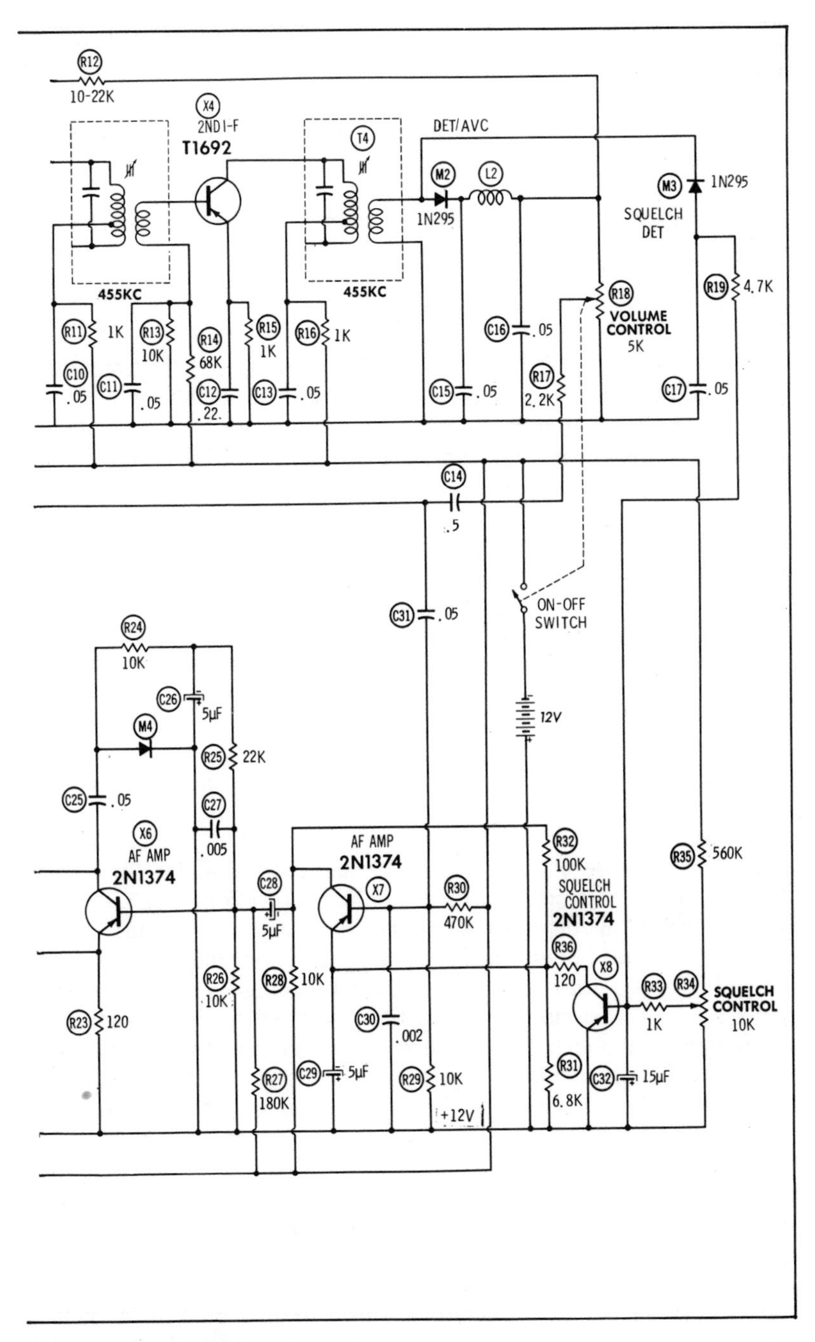

all-transistor transceiver.

Its output signal is fed back to audio-amplifier stages X7 and X6 before being coupled to X5, the crystal-controlled transmitter oscillator circuit.

Despite the miniature components and printed-circuit wiring used in transistorized equipment, circuit operation is basically the same as in tube-type equipment. Currently, about two-thirds of all CB base, mobile, and portable equipment is of solid-state construction. Examples were shown previously in Chapter 2.

KITS

The do-it-yourself fad undoubtedly accounts for the tremendous upswing in electronic kit sales, and Citizens band gear is no exception. A number of reliable manufacturers are producing some very efficient transceivers that can be easily assembled by persons with little or no previous experience.

Construction

Some kits are comprised of individual components which must be properly soldered onto printed circuit boards or wired together to form the completed circuit. Other manufacturers offer kits composed of printed-board subassemblies (receiver and transmitter oscillators, i-f, audio sections, etc.) which have been prewired, adjusted, and tested at the factory. All the builder has to do is to mount and interconnect the subassemblies. The experience derived from either type of construction should prove beneficial to the builder, especially if he is thinking of more complicated future projects.

Practically all kit manufacturers provide some type of simplified diagram to guide even the most inexperienced builder through the step-by-step construction with a minimum of effort. Fig. 3-5 shows the simplified method used by one manufacturer to illustrate how and where to mount various components on the printed circuit board. This is followed by a pictorial diagram (Fig. 3-6) which shows how the circuit board, in turn, is wired to other components mounted on the chassis. A complete schematic of the finished transceiver circuit is included with most kits, to be used in conjunction with the pictorial diagrams. When it becomes necessary to wire a component having a number of contacts, such as a rotary switch, a detailed drawing similar to that of Fig. 3-7 is provided to eliminate any confusion as to where proper connections should be made. Even the length of the wires between connection points is given in the instructions. This is important in equipment operating at high frequencies, because even a short wire represents an appreciable amount of inductance. Therefore, wiring should be kept as near as possible to the length specified in the instructions. Improper lead

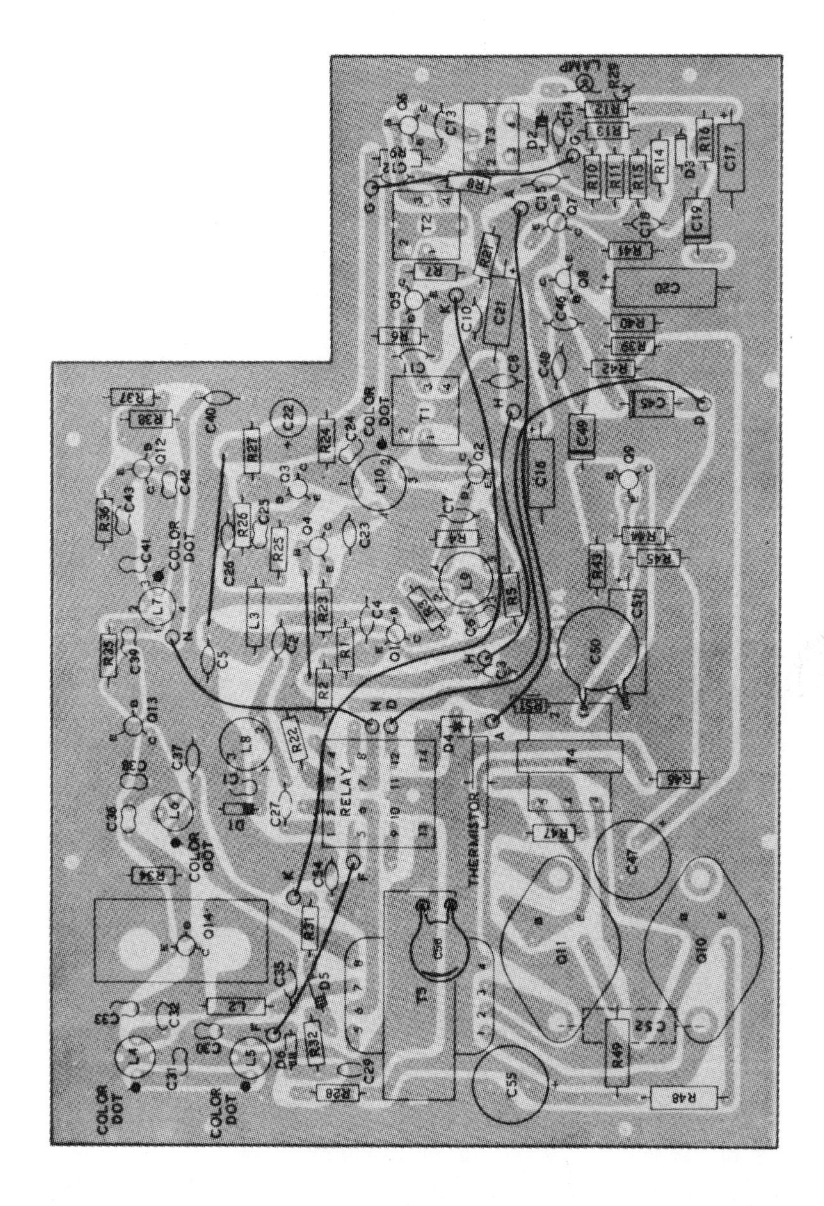

Courtesy Heath Co.

Fig. 3-5. Example of kit assembly drawing showing where to mount components on printed circuit board.

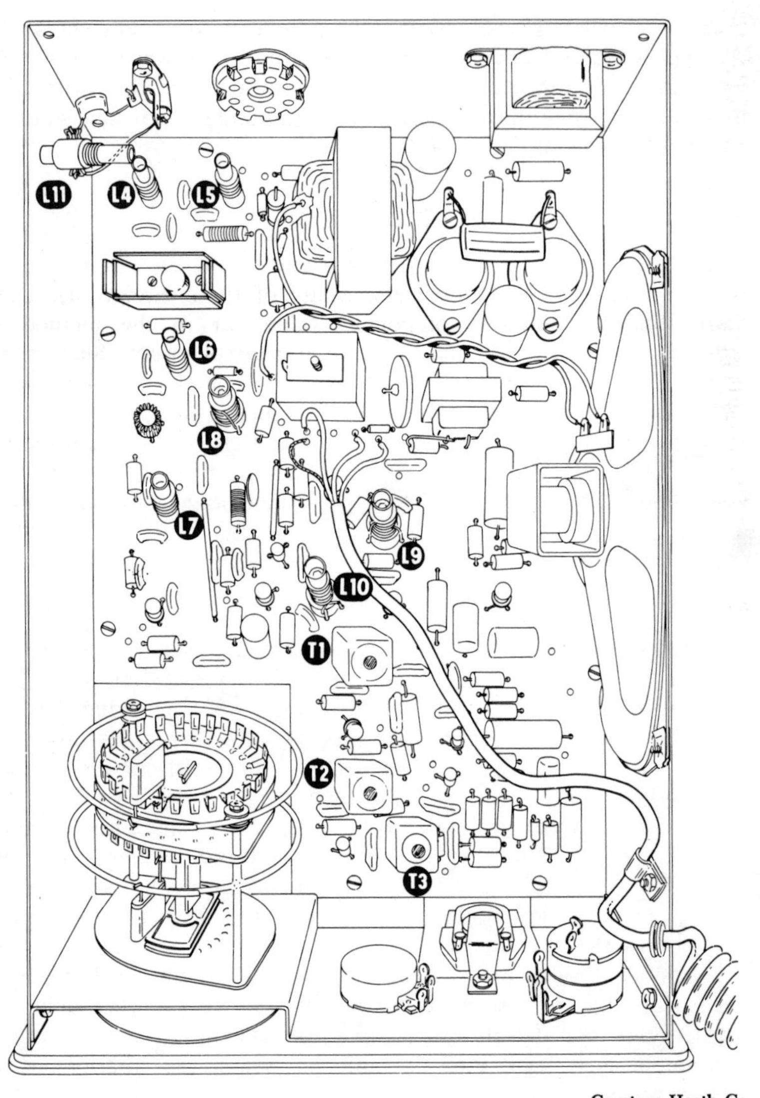

Courtesy Heath Co.

Fig. 3-6. Typical pictorial diagram used in kit construction.

dress or component placement and excessively long leads are just a few of the things to avoid during assembly.

Another pitfall that seems to be prevalent—especially among inexperienced builders—is poor soldering. Fig. 3-8 shows several types of soldered joints—obviously, only one is desirable. Too much solder on a connection is not only a waste, but may also result in a short circuit. Likewise, a cold-solder-joint or one with insufficient solder can pull loose and cause intermittent operation. However, by closely following the construction procedure outlined in the instruction manual, you should have little trouble assembling a unit that will provide many hours of dependable operation.

RECEIVER CIRCUITS

This discussion is concerned with some of the circuits in the receiver section of CB transceivers. Many of these circuits are standard, differing only in component values or in the method of coupling to other stages, and hence need no explanation. Such stages as the mixer, i-f amplifier, and detector are usually of conventional design and will not receive special attention in this chapter. By referring to the superheterodyne schematic in Fig. 3-2, you can see the type of circuitry usually employed in these stages. You will find variations from unit to unit—perhaps a semiconductor diode instead of vacuum tube in the detector stage—but the theory of operation remains basically the same.

RF Amplifier

It is highly desirable to amplify the antenna input signal before it is fed to the converter or mixer stage. The conversion stage provides little or no signal gain; moreover, the noise introduced by its operation (particularly in vacuum-tube circuits) is considerable. The rf amplifier provides the gain needed to override the inherent noise level of the converter stage, thereby increasing the signal-to-noise ratio. Both triode and pentode tubes are used as rf amplifiers in tube-type equipment. There are advantages and disadvantages to both tubes. The triode does not provide as much gain as the pentode, but it has a relatively low noise factor. The pentode provides more gain, but its noise factor is higher.

In the superregenerative receiver, a properly neutralized rf amplifier is desirable because it acts as a buffer between the antenna and superregenerative detector stage. If it were not used, detector radiation might feed back into the antenna and cause interference in other receivers.

Fig. 3-9A illustrates a typical pentode rf amplifier stage. The input signal is coupled through L1 to the grid, which is returned through R1 to the avc line. Voltage from this line varies the bias and subsequent gain of the stage in proportion to the carrier level of the incoming station signal. The amplified signal at the plate of this tube is developed across T1, which serves as both a plate load and a coupling device. T1 is tuned to respond only to a desired range of frequencies; that is, it acts as a high impedance only to the signals for which it is tuned. The desired signals are dropped across the primary the same as if it were a load resistor. These signals are then inductively coupled through the secondary to the converter stage.

A transistor rf amplifier stage is shown in Fig. 3-9B. This circuit, unlike the previous example, is inductively coupled to the antenna through T1 instead of by the direct method in Fig. 3-9A.

Another popular rf amplifier circuit is shown in Fig. 3-9C. In this circuit a 6DS4 triode nuvistor is coupled directly to the antenna. Because of its compact design and extremely low-noise figure, the nuvistor is ideally suited for use as an rf amplifier. This circuit must be neutralized, however, before it will operate properly. Capacitor C1 provides the means of achieving neutralization.

Oscillators

Most oscillators in CB equipment are crystal-controlled, especially in transmitter circuits. You will find receivers that are not crystal-

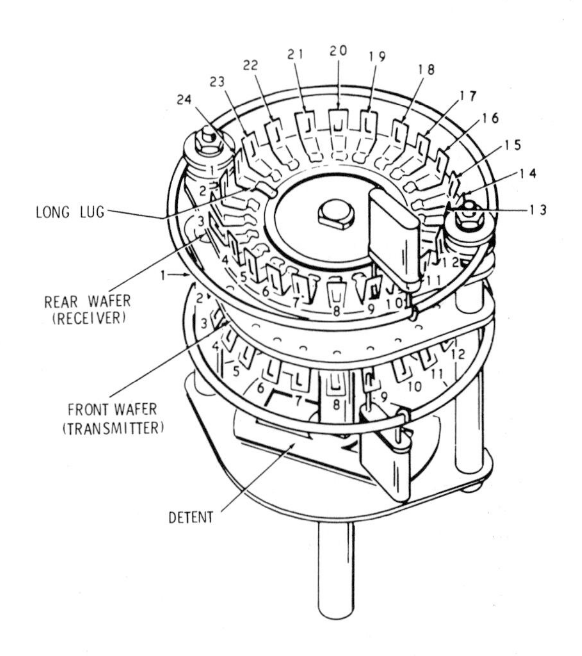

Fig. 3-7. Complex switch connections are

controlled, and others that are controlled only on some channels and tunable over the remainder of the band.

Crystals—Before getting into the operation of the crystal-controlled oscillator, it is desirable to have a basic understanding of what crystals are and how they control the oscillator frequency.

A number of years ago, early experimenters discovered that certain crystalline substances, in their natural state, produce an electrical charge when subjected to mechanical strain. Later it was found that the opposite effect also held true—a small plate properly cut from this crystal and placed between two electrodes will exhibit a mechanical strain when a voltage is applied to the electrodes. This is known as piezoelectric effect, meaning "pressure electricity."

There are a number of materials having such properties, two of which are Rochelle salt and quartz. The latter crystal has been proven superior in performance and is presently the most popular in this application.

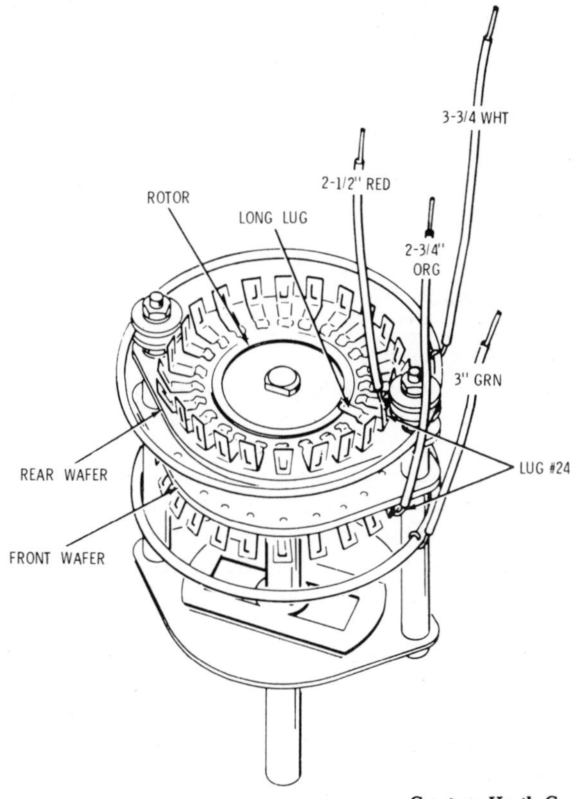

Courtesy Heath Co.

detailed to guide builder.

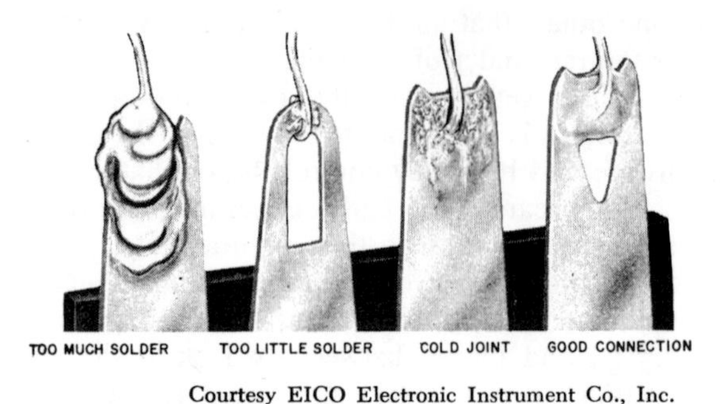

Courtesy EICO Electronic Instrument Co., Inc.

Fig. 3-8. Properly soldered connections are a must in kit construction.

Quartz, which is silicon dioxide in its natural state, is found in hard crystallized six-sided prisms resembling glass (Fig. 3-10). Each crystal has three axes—X, Y, and Z. The angle and direction at which the crystal slabs are cut from the raw stone help determine the characteristics of the finished crystal. After being cut and ground to the proper specifications, the crystal is mounted in a holder. Fig. 3-11 shows the type of crystal normally used in CB equipment.

Fixed-Tuned Oscillator—The operation of an rf generator (oscillator) can be precisely controlled by using a crystal as the frequency-controlling element. Fig. 3-12A shows a crystal-controlled Colpitts oscillator with frequency doubling in the plate stage. Half of a 6U8 tube is used as the oscillator; the other half (not shown) serves as the mixer. The oscillator output is coupled to the mixer section of the 6U8 tube via the interelement capacitance of the tube socket. The crystal (M1) in the grid circuit determines the oscillating frequency of the circuit.

Continuously Tunable Oscillator—Some receiver oscillators are not crystal-controlled. Instead, they may use a continuously variable tuned-oscillator circuit like the one in Fig. 3-12B. This circuit, like that of Fig. 3-12A, employs the triode section of a 6U8. You will notice, however, that a tank circuit consisting of L1, C4, C5, and C6 is used in the oscillator grid circuit in place of a crystal. The resonant frequency of this tank, and the subsequent rf output applied to the mixer, can be varied by changing the setting of receiver tuning control C6.

Combination Fixed-Tuned and Tunable—Fig. 3-12C shows a combination oscillator circuit employing half of a 6AN8 tube. When switch M1 is in position 3, the oscillator is fixed-tuned at a frequency determined by crystal M2. When the switch is in position 2, crystal M3 is connected into the grid circuit and determines the oscillator frequency. In position 1, an adjustable tank circuit comprised of C3, C4, and L1 is connected into the grid circuit, thus providing a

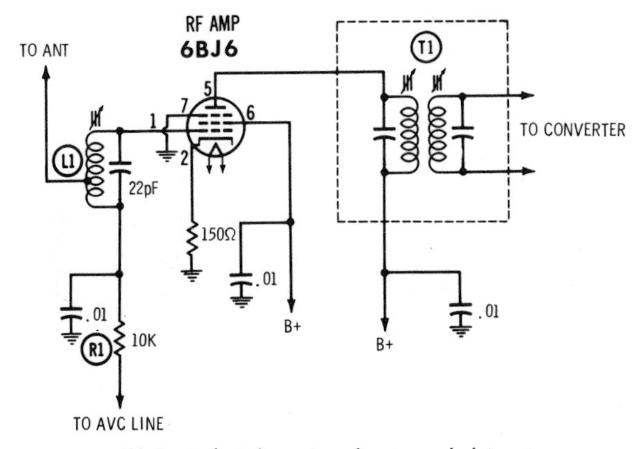

(A) Pentode tube using direct-coupled input.

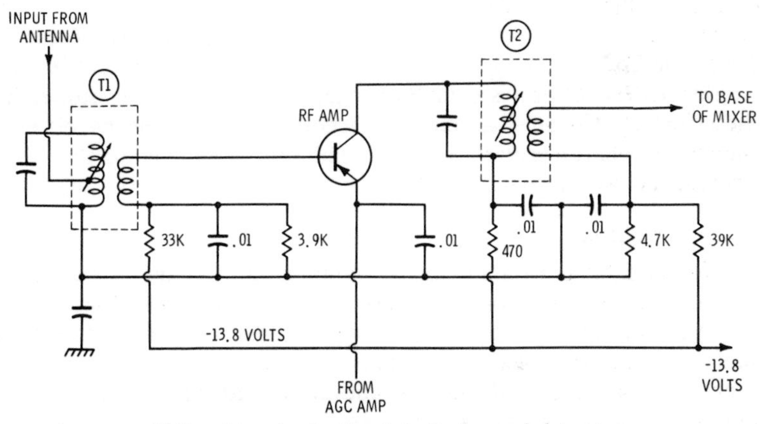

(B) Transistor circuit using inductively coupled input.

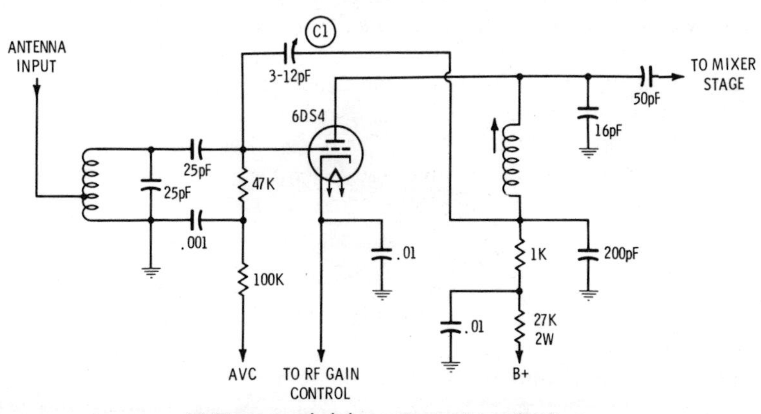

(C) Direct-coupled, low-noise nuvistor circuit.

Fig. 3-9. Typical rf amplifiers employed in CB transceivers.

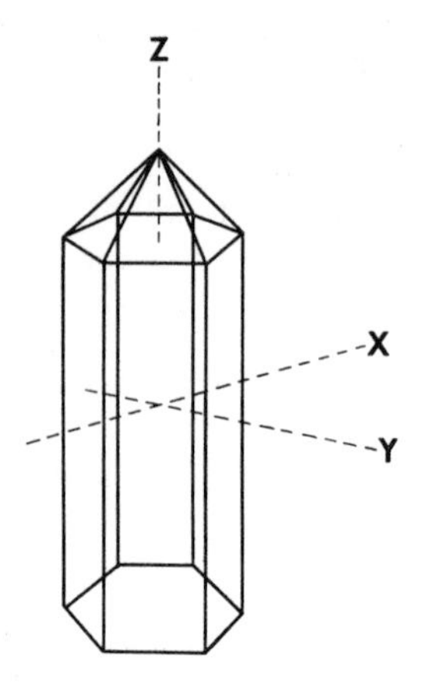

Fig. 3-10. The quartz crystal in its natural state.

tunable oscillator. The frequency is changed by means of tuning capacitor C3.

Fig. 3-12D illustrates a multichannel, fixed-tuned, transistor oscillator circuit. Channel selection is accomplished by means of SW1 and the resultant oscillator output is inductively coupled to the mixer stage through T1.

Audio Amplifiers

The circuit requirements for the audio-amplifier section of the CB transceiver are far from exacting. Most commercial broadcasting stations transmit a relatively high-fidelity signal that occupies a wide band of frequencies. In order to reproduce the transmitted signal with a high degree of quality, the broadcast receiver must also employ circuitry with wide-response characteristics. However, for strictly voice transmissions, such high-quality response is not needed and would only take up unnecessary space in the frequency spec-

Fig. 3-11. Example of the type of crystal employed in Citizens-band equipment.

trum. Therefore, the FCC has limited the amount of space that can be occupied by the emission from a CB transmitter as stated in Chapter 2.

Generally, you will find that a conventional two-stage audio section comparable to that in Fig. 3-13 is used in units with no squelch circuit. The audio section shown here is straightforward, using half of a 6U8 (V1A) as a voltage amplifier driving a 6AQ5 pentode power-output stage. Transformer T1 serves as both audio and modulation transformer. The audio-amplifier section of a receiver employing a squelch circuit will differ from the conventional arrangement just shown; this will be covered when squelch circuits are discussed.

Single-ended output stages are used in most transceivers, although you will also find some of the double-ended (push-pull) variety. Fig. 3-14 shows a push-pull audio system from an all-transistor unit.

A 2N238 pnp transistor serves as a voltage amplifier, or driver, for the output stage. It is connected in a common-emitter configuration, and the bias is set by a voltage divider comprised of R1 and R2. Notice that the audio signal is fed to volume control R1 through the movable arm, instead of by the usual manner in which a portion of the signal voltage dropped across the resistance is tapped off and fed to the amplifier. The arrangement shown here allows the resistance of R1 to remain constant regardless of the position of the movable contact. Hence, the resistance offered the audio input signal will vary with the control setting, while the total resistance across R1 remains constant in order for it to function as part of the bias network.

The dc operating point of the emitter is determined by the value of R4, and emitter bypass capacitor C3 is used to prevent degenertion. Collector voltage for transistor X1 is obtained from the −9.6-volt source through the primary of driver transformer T1.

The output stage employs a matched pair of 2N185 transistors (X2 and X3) connected in a push-pull arrangement. The base bias, determined by the combined value of R5 and R6, is applied through the secondary of driver transformer T1. Thermistor R6 serves as a means of temperature compensation to stabilize circuit operation over a wide ambient-temperature range with a minimum of distortion. R7, a common emitter resistor for X2 and X3, compensates for circuit or output variations.

Automatic Volume Control

Automatic volume control (avc) provides a means of controlling the receiver gain so it will vary in inverse proportion to the strength of the incoming signal. This is a desirable circuit feature because it maintains the receiver output at a relatively constant level as the input signal varies. In other words, if you were receiving a transmission and the strength of the signal entering the antenna dropped, the avc circuit would automatically increase the gain of certain stages within the receiver to compensate for the loss. In this way a constant volume level is maintained at the speaker. Such decreases in signal strength are likely to occur on weak signals, for example, when a unit is transmitting from a moving vehicle in an area where buildings and other obstructions are blocking the signal.

Usually the rf amplifier and one or more i-f stages are controlled by the avc circuit. The more stages under control, the more efficient is the avc action. It is desirable that at least two stages of the receiver be controlled.

The avc voltage can be derived from either the second-detector diode or a separate avc rectifier. Fig. 3-15 shows an automatic volume-control circuit using the first method. Here M1 serves as both the

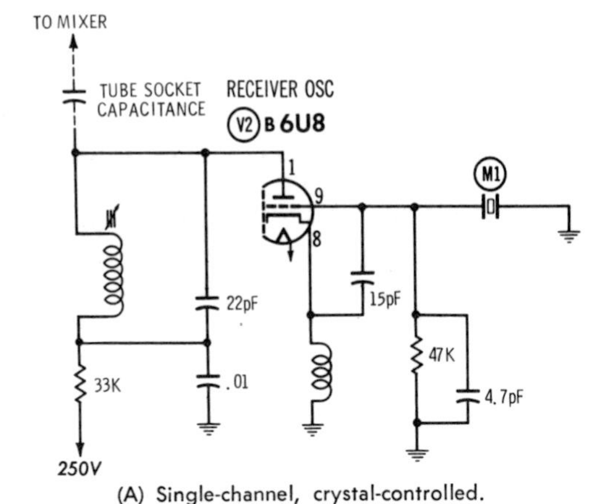

(A) Single-channel, crystal-controlled.

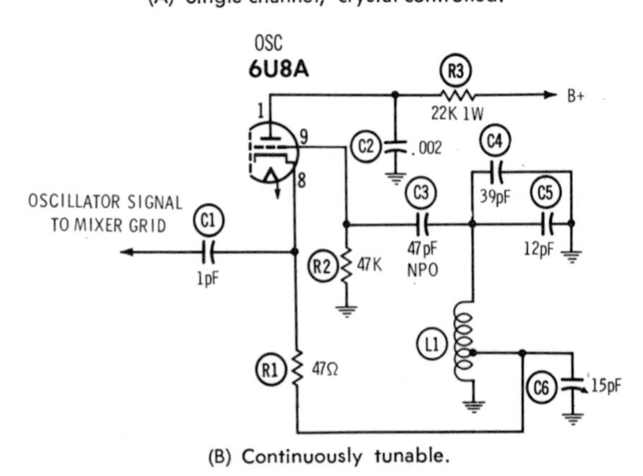

(B) Continuously tunable.

Fig. 3-12. Typical oscillator

second detector and avc diode. In this circuit the diode is connected so that the negative half of the signal is detected. By doing so, the diode can also be used to produce the desired avc voltage. The audio signal, available at point *A*, is passed on through succeeding circuitry to the next stage. R5 is the avc load resistor; it and C5 form an avc filter to remove the ac component. At the output of this network, a negative dc voltage varies in accordance with the average carrier level of the station being received. The avc voltage derived from this circuit is fed to the grids of the rf amplifier and second mixer stages to vary the bias and subsequently the gain of the stage.

Circuits using a separate avc diode will be somewhat similar to Fig. 3-16. Here, a portion of the i-f signal is tapped off at the secondary of transformer T2 and applied to diode M1. Variations in the rectified signal are filtered by C21 and R26 before the avc voltage is applied to the controlled stages.

One variation of the automatic volume-control circuit is delayed

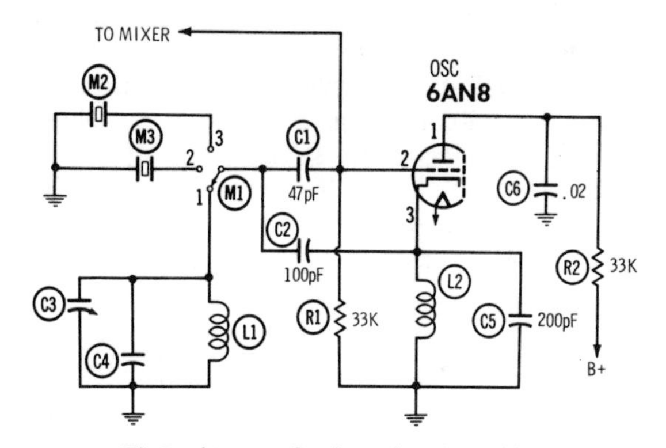

(C) Combination fixed-tuned and tunable.

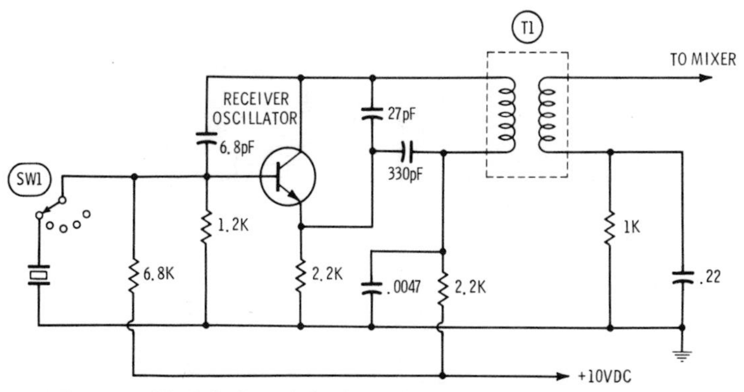

(D) Multichannel fixed-tuned transistor oscillator.

circuits in CB receivers.

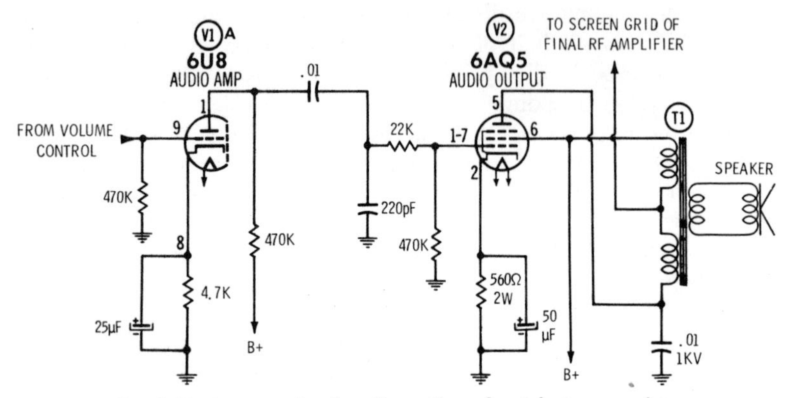

Fig. 3-13. A conventional audio section of a tube-type receiver.

avc. It stands to reason that an avc control voltage is not needed on the weaker signals. In fact, the full sensitivity of the receiver must be utilized to provide sufficient amplification. In the previous arrangement, some signal attenuation will occur even on weak signals. With the delayed avc circuit, however, the control voltage is not developed until the incoming signal strength is above a certain level. In this way the weaker signals will receive a full amplification. The method used to delay the circuit action is really quite simple and merely involves applying a dc voltage in series with the avc diode-

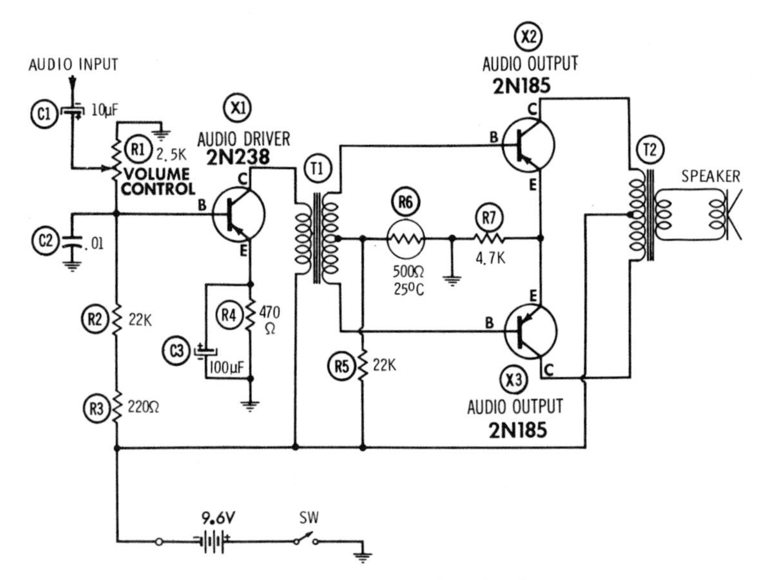

Fig. 3-14. A typical transistorized push-pull audio-output stage.

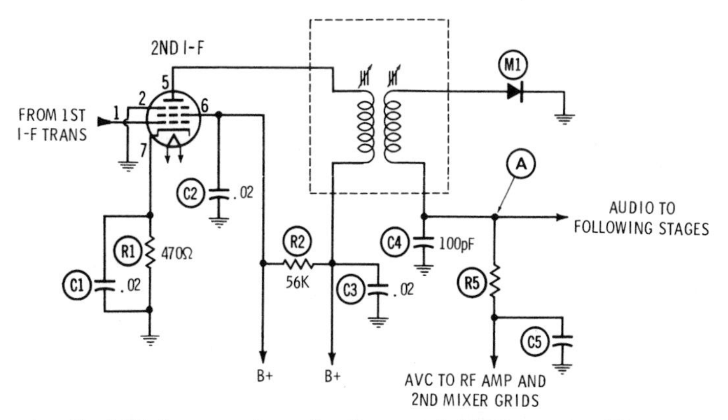

Fig. 3-15. An avc system using the second detector as a rectifier.

load resistor. This voltage is applied in such a way that the plate of the diode is maintained more negative than the cathode while no signal is being received, thereby preventing the tube from conducting. Should a weak signal be received, it would obtain the full amplification since no avc voltage is being produced. If an incoming signal produces an rf voltage higher than the delay voltage, the plate of the diode will be driven positive and cause the tube to conduct, thereby producing the avc voltage.

A separate avc diode must be used with a delayed circuit of this type. Using the audio detector diode would prevent demodulation of the weaker signals and cause distortion of the audio signal.

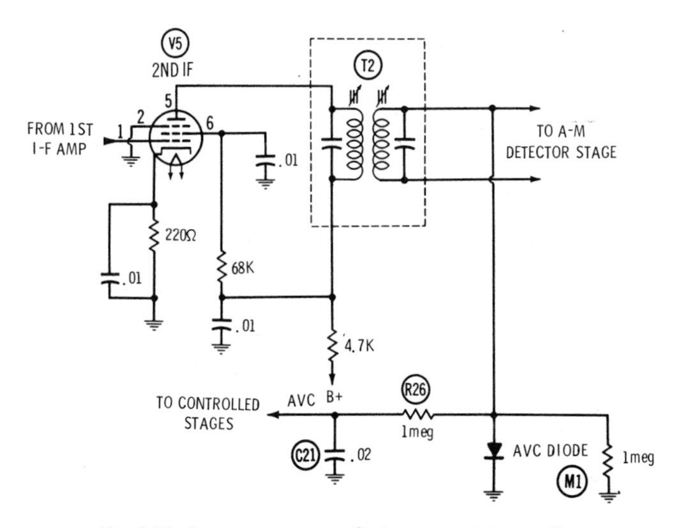

Fig. 3-16. An avc system employing a separate rectifier.

Squelch

The purpose of the squelch circuit is to block the speaker output when no signal is being received. This feature is included in the majority of Citizens band equipment. Without it, background noise can be heard from the speaker at all times except when a signal of sufficient magnitude is being received. A squelch circuit is desirable to prevent undue listener fatigue.

Two types of squelch circuits are used. One is the signal-operated type that depends on the strength of the incoming signal; the other is the noise-operated circuit that utilizes the background noise present during no-signal conditions. The former type is by far the most common in CB applications.

A typical signal-operated squelch circuit is shown in Fig. 3-17. Here, audio amplifier V2B is disabled by the squelch-control voltage. This occurs when the received carrier amplitude falls below a cer-

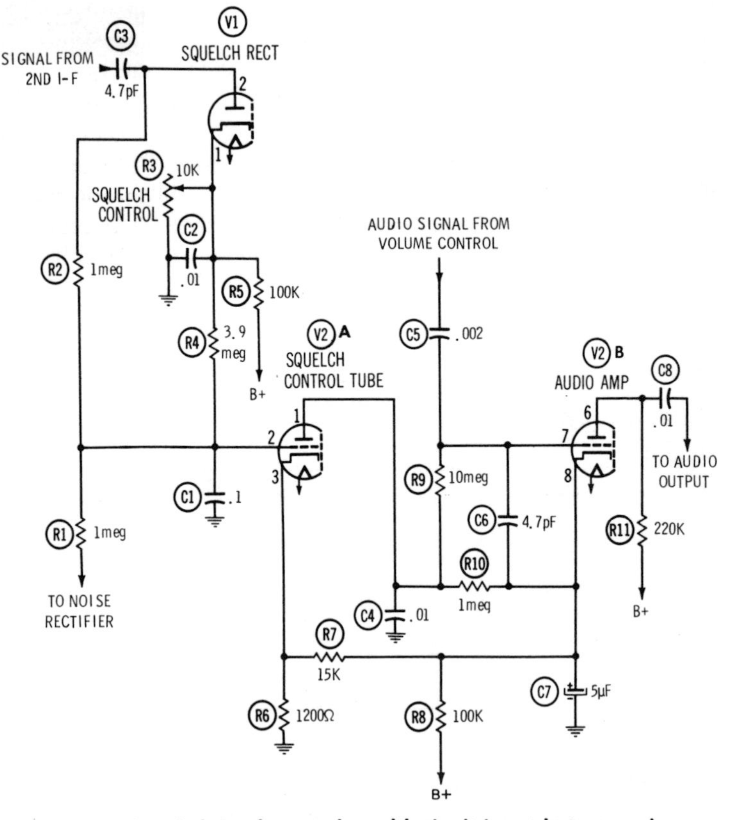

Fig. 3-17. A typical signal-operated squelch circuit in a tube-type receiver.

tain level, as determined by the setting of squelch control R3. A portion of the rf signal is taken off at the output of the second i-f amplifier stage and coupled through C3 to squelch rectifier V1. This tube rectifies the carrier signal, thereby developing a negative voltage which varies according to the strength of the carrier signal. This voltage is then fed to the grid of the squelch-control tube, where it opposes the positive voltage applied to this element by R4. When a normal signal is being received, the total voltage to ground is sufficiently negative to prevent V2A from conducting. Under this condition, audio amplifier V2B operates as a conventional RC coupled voltage-amplifier stage. Bias is obtained as a result of the contact potential current flowing through R9. Should the incoming carrier signal be removed or its level drop below the value designated at the setting of R3, the bias for V2A will become more positive, thereby allowing V2A to conduct. The resultant plate current of squelch-control tube V2A flows to B+ through resistors R10 and R8, producing across R10 a voltage drop that biases audio amplifier V2B beyond cutoff. This of course blocks (squelches) the audio signal and nothing will be heard from the speaker until the arrival of a signal strong enough to overcome the squelch voltage.

A transistor squelch circuit is shown in Fig. 3-18. In this circuit, squelch-control potentiometer R1 changes the state of Q1 causing this transistor to conduct either more or less current. The amount of conduction through Q1 determines the emitter voltage at the first audio amplifier. When the setting of R1 is such that Q1 is conducting and the first audio transistor is cut off, the audio will be disabled and no sound will be heard from the speaker. When an rf signal is received, however, it causes Q1 to cut off (depending on the setting of R1) and the first audio amplifier is again forward biased. Thus, the audio amplifier begins to conduct and the audio is heard.

Noise Limiters

The majority of CB equipment employs some type of noise-limiting circuit. Its purpose is to prevent noise pulses from reaching the

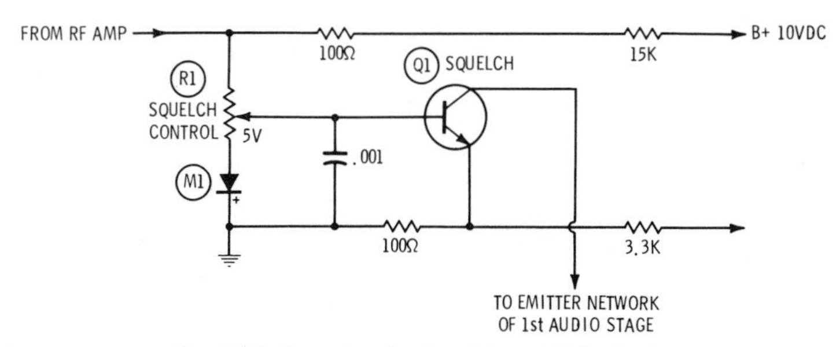

Fig. 3-18. Example of a transistor squelch circuit.

audio amplifier and causing objectionable interference. This circuit can be a simple arrangement or it can be complex. Either one will perform the same job, but its efficiency is influenced to a great extent by circuit design.

Automobile ignition systems are just one source of noise-pulse interference you will encounter. Atmospheric disturbances and devices such as electric fans, drills, sweepers, mixers, and relay contacts are common sources, to name just a few. These pulses affect the amplitude of the radio signal; hence, they are especially troublesome in a-m equipment. The fm signal, even in the presence of such interference, remains relatively unaffected by it. This is due primarily to the action of the limiter circuit(s), which clip amplitude variations from the signal.

The noise limiter reacts to the interfering pulses by momentarily blocking the signal during the time the interfering pulses occur. So instantaneous is this action that it will be unnoticed. Some transceivers have provision for switching the noise limiter out of the circuit when the existing interference is not strong enough to be objectionable.

Fig. 3-19 shows a noise-limiter circuit. The output from diode detector V4B is applied through R28 to the plate of V4A. The voltage at this element is such that V4A conducts when normal signals are received. When noise pulses cause the audio peaks to exceed a certain negative value, the plate will become more negative than the cathode and the tube will cease to conduct, thereby blocking both

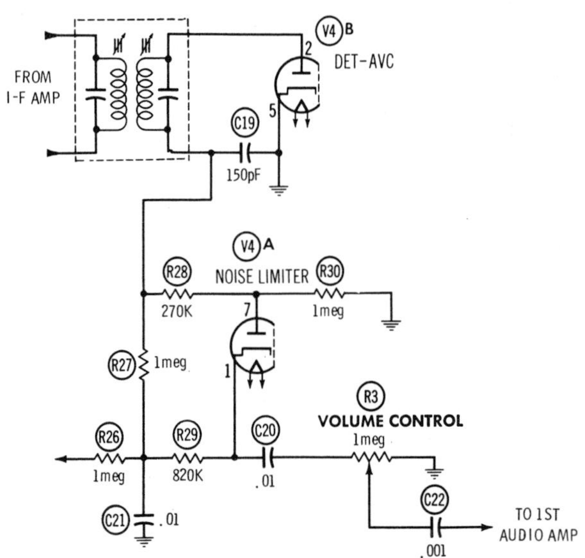

Fig. 3-19. Example of a noise-limiting circuit using a vacuum-tube diode.

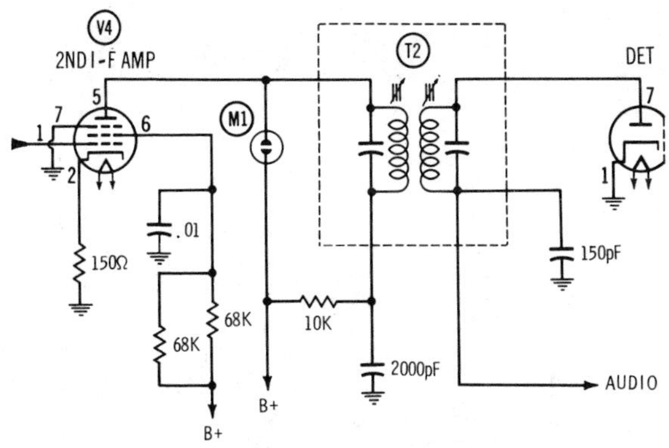

Fig. 3-20. A simple noise-clipping circuit using a neon bulb.

the audio signal and the noise pulse at that particular instant. The junction of R26 and R29 is bypassed for audio by C21 and serves as a reference for bias of the diode. The time constant here is short enough that the bias will follow changes in the avc voltage. The audio signal passed by this stage is developed across volume control R3 and fed to the first audio amplifier.

A vacuum tube does not always have to be employed in a circuit of this type. Fig. 3-20 illustrates a simple method of blocking the unwanted noise pulses with a neon lamp. This lamp (M1) is connected from the plate of V4 to B+; therefore the voltage developed across the primary of T2 will be applied to it. When the signal is normal, the voltage applied to M1 will not be sufficient to trigger it. However, if a noise pulse of sufficient amplitude is dropped across the primary of T2, it will develop enough voltage to ionize M1, thereby shorting the primary of T2 and subsequently blocking both the signal and the pulse at that instant.

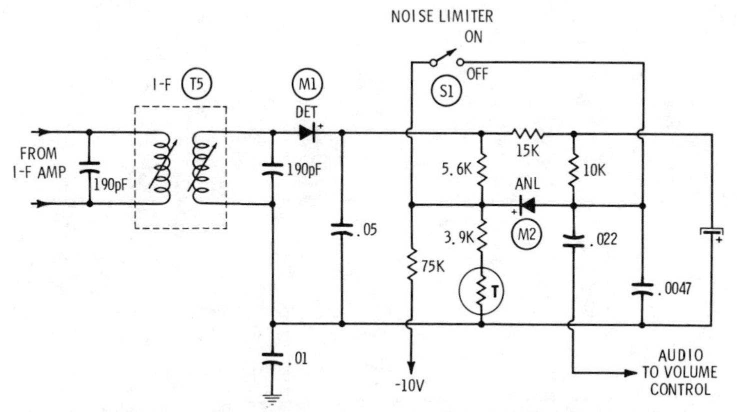

Fg. 3-21. A series-type noise limiter using a solid-state diode.

Fig. 3-21 shows still another configuration, this one using a solid-state diode (M2) for limiting purposes. This circuit forms a series-type noise limiter which provides i-f clipping to reduce noise peaks. When switch S1 is open, diode M2 is effective and limiting occurs; in the closed (OFF) position, the switch shorts out M2, thereby disabling the circuit.

TRANSMITTER CIRCUITS

Transmitter circuits are relatively simple in comparison with those of the receiver. In fact, a single rf oscillator stage properly connected to an antenna is capable of radiating an rf carrier into space. Additional stages are needed, of course, if any modulation is to be impressed on this signal. CB transmitters are generally composed of four basic stages—speech amplifier, modulator, rf oscillator, and rf power output. Some also include a microphone preamp ahead of the speech amplifier. Fig. 3-22 shows a functional block diagram of a typical transmitter using these stages. Some do not include the final rf power-output circuit, in which case the oscillator stage feeds the antenna. The latter then becomes the final stage, and the modulator output is fed to it.

Speech Amplifier

A separate speech-amplifier circuit can be used, but normally the audio amplifier, or driver, of the receiver doubles as a speech amplifier when transmitting. This stage employs conventional circuitry in practically all transceivers. It serves as a voltage amplifier to provide an audio-signal output of sufficient magnitude to drive the power-output stage. Now, however, instead of amplifying the receiver signal, a small audio voltage produced by the microphone is fed into its input circuit.

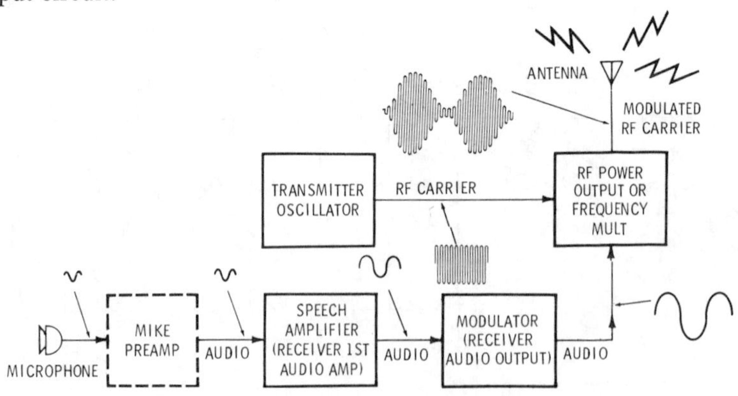

Fig. 3-22. Functional block diagram of a typical Class-D transmitter.

Fig. 3-23 is a typical example of a receiver audio section designed to function as part of the transmitter. In this circuit the microphone output is first coupled through L6 to the grid of V4B, a microphone preamp stage. Here it receives an initial boost in strength before being coupled through C22 to the input of speech amplifier V3B. Theoretically, the microphone preamp could be classed as a speech amplifier since it also precedes the modulator.

Modulator

Continuing with the audio section shown in Fig. 3-23, we see that after passing through the voltage amplifier, the signal is RC coupled to a conventional audio power-output stage which now serves as the modulator. Present at the output is the original audio signal from the microphone, amplified many times above its original value and of sufficient power to perform its intended function of varying the rf carrier level in accordance with the signal from the microphone. Both the speech amplifier and the modulator circuits operate in a straightforward manner, treating the incoming microphone signal as if it were coming from the detector stage of the receiver. Here again a separate tube or transistor can be used as the modulator, but you will find that most units draft the existing audio-output stage to perform a dual function. In this circuit, the primary of output trans-

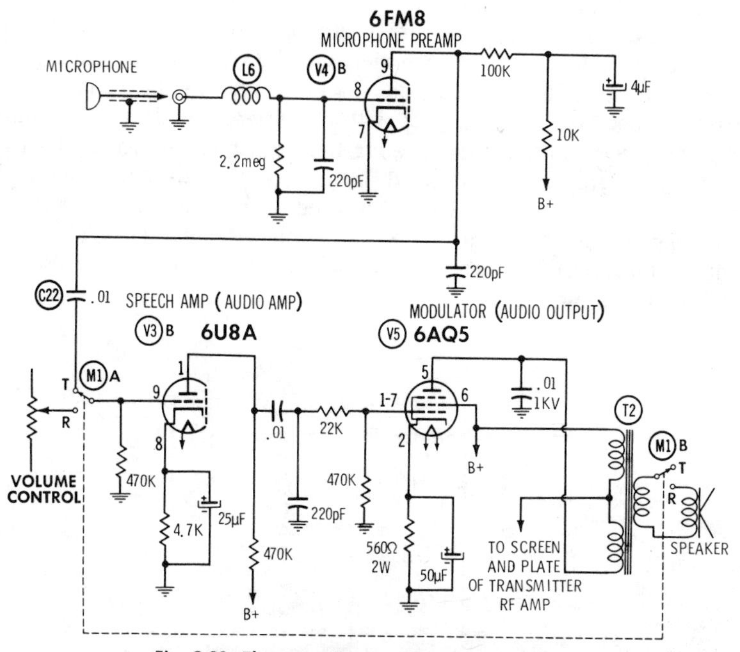

Fig. 3-23. The transmitter audio amplifier section.

former T2 is center tapped to permit it to serve also as an autotransformer for modulation purposes. The audio-output voltage from V5 is developed across the primary and fed through the center tap to the plate and screen of a pentode rf power amplifier.

Oscillator

The audio signal from the microphone could be passed through stage after stage of amplification until it reached great proportions—yet if it were applied to the transmitting antenna, it would not be radiated into space like the station signal. Why is this so? Simply because it lacks an rf carrier. Signals in the rf (radio-frequency) range have the ability to radiate from a conductor into free space. Since the audio signal is the one desired at the receiving station, it is superimposed on the rf signal before the latter is transmitted from the antenna. Hence, the rf signal is known as the carrier. At the receiver the two are separated and the carrier is disposed of, while the microphone signal is amplified and then reproduced by the speaker.

The purpose of the transmitter oscillator is to generate the rf carrier signal. Fig. 3-24 shows the circuit of a crystal-controlled oscillator for a Class-D transmitter.

This circuit employs the triode section of a 6U8A tube as an rf oscillator, the frequency of which is controlled by crystal M1 in the grid circuit. Tunable inductor L6, together with its parallel stray capacity, forms a tuned plate load. An optimum load for third-harmonic operation on all CB channels can be obtained by tuning L6 to resonate with its stray capacity at the center of the Class-D frequency range. C11, in parallel with the crystal, adds to the shunt capacity across it, thereby providing the proper crystal load. Positive feedback to the grid is derived from the voltage-divider network comprised of C12 and C10, and is used to increase the amount of drive. The rf output from this circuit is then fed to the grid of the final rf amplifier. The transistor counterpart to the tube-type oscillator circuit is shown in Fig. 3-25.

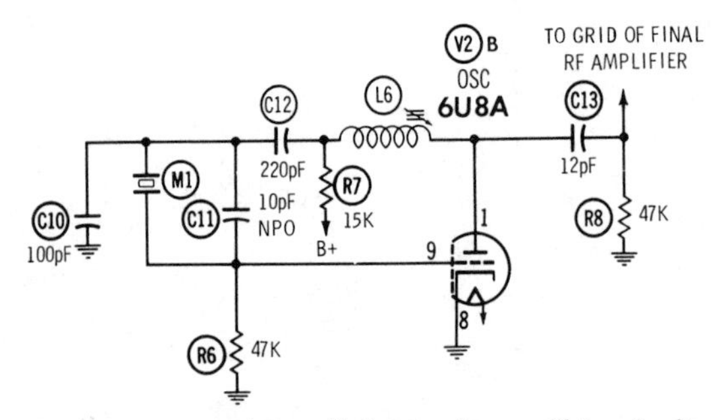

Fig. 3-24. A crystal-controlled transmitter oscillator circuit.

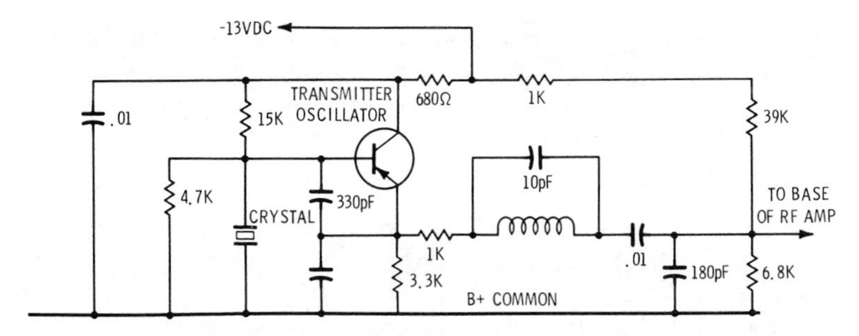

Fig. 3-25. Example of a transistorized crystal-controlled transmitter oscillator.

Another interesting oscillator arrangement is shown in Fig. 3-26. In this circuit, a dual purpose 6BH8 tube is used for the transmitter and receiver oscillators. The triode section functions as a five-channel crystal-controlled oscillator for the receiver, and its output is coupled to a separate mixer tube. Unlike in the previous circuit, a pentode (the second half of V3) instead of a triode is employed as the trans-

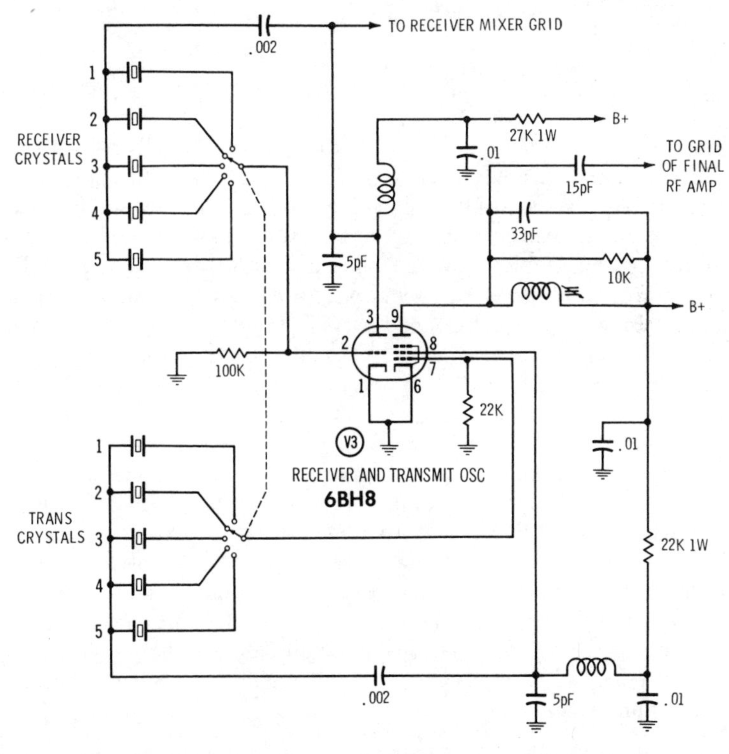

Fig. 3-26. A circuit incorporating the transmitter and receiver oscillators in the same tube.

mitter oscillator. It, too, is crystal-controlled for five-channel operation. A ganged switching arrangement is used for simultaneous selection of the desired transmitter and receiver crystals. The output of the transmitter oscillator is fed to a final rf amplifier.

RF Power Output

So far, we have followed the audio signal from the microphone, through the various stages of amplification, up to the output of the modulator, and we have seen how the rf carrier is produced by the oscillator. Here, however, we are concerned with the final rf amplifier stage, where both of these signals are injected to produce the modulated carrier.

Most transceivers employ some form of plate modulation, which must always occur in the final rf stage of the transmitter. As mentioned previously, not all transceivers have an rf power output; the transmitter oscillator feeds the antenna and is therefore the final rf stage. It should be pointed out that the modulator stage merely provides the proper audio signal. The modulation process itself occurs in the power-output stage.

Oscillators

Fig. 3-27 shows a circuit using a popular method of plate modulation. Here the rf oscillator signal is injected through C26 into the grid of the rf power-output tube. Audio modulator V3, operating as a conventional Class-A power-output stage, develops the desired audio signal across the primary of T1. This transformer serves as the audio output when receiving, and as an autotransformer for the modulation function. Notice how the plate and screen of V4 receive B+. The plate of modulator V3 is supplied through the primary of T1, whereas V4 is supplied from the primary center tap. Now, with the transmit button depressed and no sound entering the microphone, the carrier generated by the oscillator will be amplified by V4 and radiated from the antenna in its original, unmodulated form. The B+ supplied to V4 will be constant.

When an audio signal is introduced, the varying plate current of V3 will develop a corresponding voltage drop across the primary of T1. This will alternately add to and subtract from the plate and screen voltages of V4, causing the rf output to vary above and below its normal level in accordance with the audio signal.

In this circuit, the screen and plate of the final tube are both used to modulate the carrier. This is generally done when pentodes or beam tetrodes are used, because of the higher percentage of modulation obtained.

You will also find that most rf power-output stages operate as Class-C amplifiers. The advantage here is that the plate efficiency is

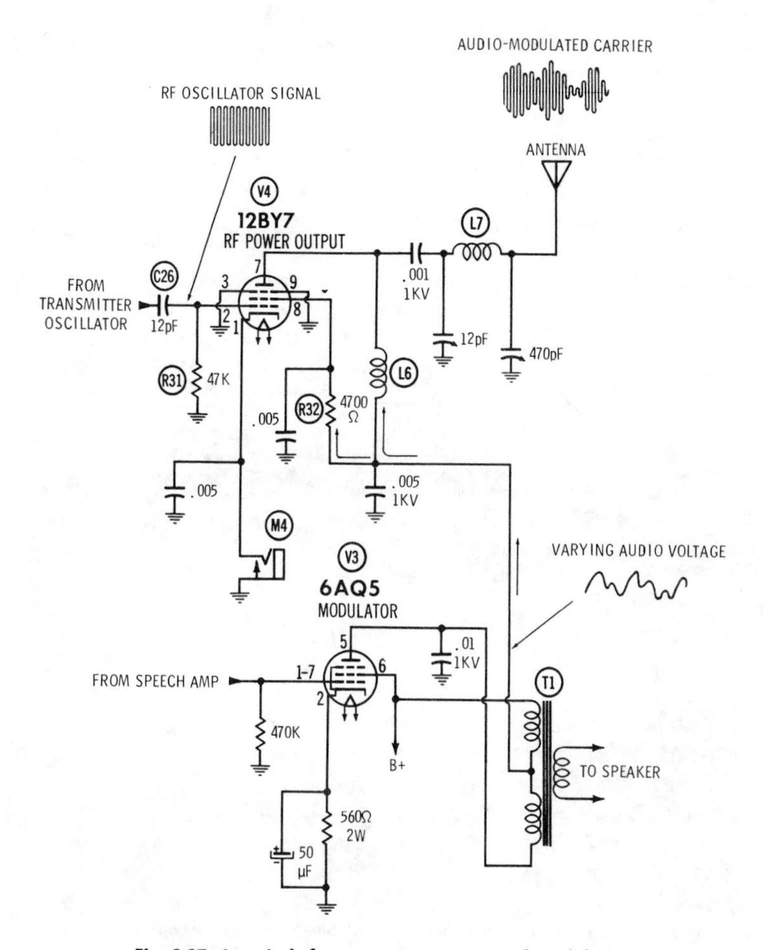

Fig. 3-27. A typical rf power output stage and modulator.

increased—in short, a higher power output is provided for a given dc plate power input.

SINGLE SIDEBAND

The congested band conditions in various radio services has dictated the need for radio equipment that will occupy less operating space in the frequency spectrum. This need has been met by the introduction of single-sideband equipment. In conventional a-m radio transmission, the carrier signal and *two* sidebands are transmitted. One sideband is present on each side of the carrier and both contain the same intelligence—the desired voice signal. From this it is obvious that the same information can be conveyed by transmitting only one of the sidebands, and only half as much space in the frequency spectrum is occupied by such a signal.

With single sideband the carrier is suppressed before transmission, so all power is concentrated within the one sideband. The absence of a carrier also eliminates interference. However, a carrier is reinserted at the receiver to permit demodulation.

A variety of single-sideband gear is currently available for CB use. One such unit is shown in Fig. 3-28. This transceiver, the General Radiotelephone Model SB-72, covers all Class-D channels and provides selection of either upper or lower sideband. The unit can also be operated in the a-m mode. A built-in universal power supply permits operation from either 115-volt ac or 12-volt dc power source.

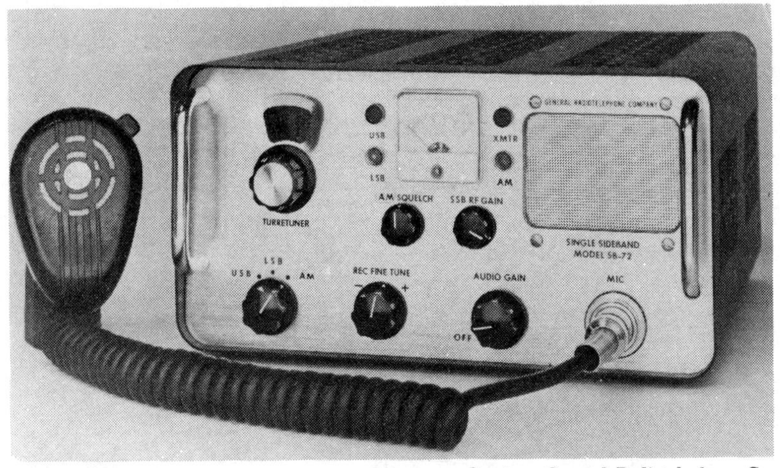

Courtesy General Radiotelephone Co.

Fig. 3-28. The General Radiotelephone Model SB-72 transceiver operates in either single-sideband or a-m modes and provides coverage of all Class-D channels.

POWER SUPPLIES

In most Citizens band equipment, you have a choice of power supplies. Equipment is available to operate on self-contained batteries, or from external 6-, 12-, or 24-volts dc, or 115- or 230-volts ac. Many units have provisions for a choice of sources.

AC Supplies

Receivers designed to operate from 115 volts ac only will usually employ a conventional power transformer and rectifier to supply the dc voltages necessary for proper operation of the various stages. Other units employ silicon or germanium rectifiers.

DC Supplies

Equipment designed strictly for operation from a dc source will include one of three types of power supplies—battery, vibrator, or transistor.

Battery—The compact, portable all-transistor phones operating at relatively small outputs use self-contained batteries exclusively. Since the current drain is light, and high dc voltages are not required for transistor operation, batteries are all that are needed to provide many hours of dependable operation.

Vibrator Supply—Larger equipment using vacuum tubes requires a higher dc voltage for operation, and will probably incorporate a vibrator power supply (Fig. 3-29). Here, we already have dc at the input, but it must be converted to a form of ac before it can be stepped up by transformer T1.

The purpose of the vibrator is to chop up the steady dc current in order to produce square-wave dc pulses in the transformer primary. The square waves appear as a high-voltage ac sine wave across the secondary, which is then applied to the rectifier and filter section of the supply. Semiconductor diodes M6 and M7 are used for rectification and, together with C4 and C5, form a voltage-doubler arrangement. The pulsating dc output from the rectifier is fed to the filter section, where it is handled in the same manner as in an ac supply. From a meager 12 volts dc at the input of this circuit, an output of approximately 250 volts dc is achieved.

Universal Supply—In addition to the single-voltage power supplies discussed, equipment can be obtained with a combination, or universal, supply that will operate from either ac or dc. Many of the

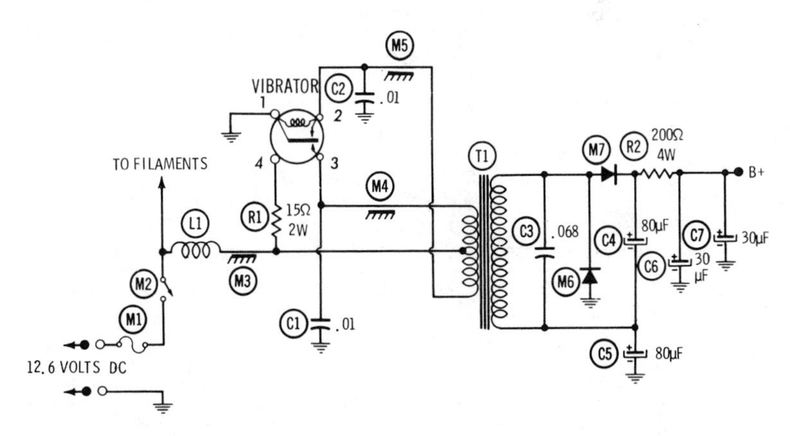

Fig. 3-29. Vibrator power supply employed for dc operation in some tube-type transceivers.

CB units on the market (some of them were shown in Chapter 2) have this feature.

The power supply shown in Fig. 3-30 operates from 117 volts ac and from both 6 and 12 volts dc. A conventional full-wave ac power supply is located at the bottom of the schematic. Directly above it is a vibrator circuit that is common for both 6- and 12-volt dc operation. A single multielement socket on the transceiver accommodates all three inputs, although individual plugs (Fig. 3-31) are used be-

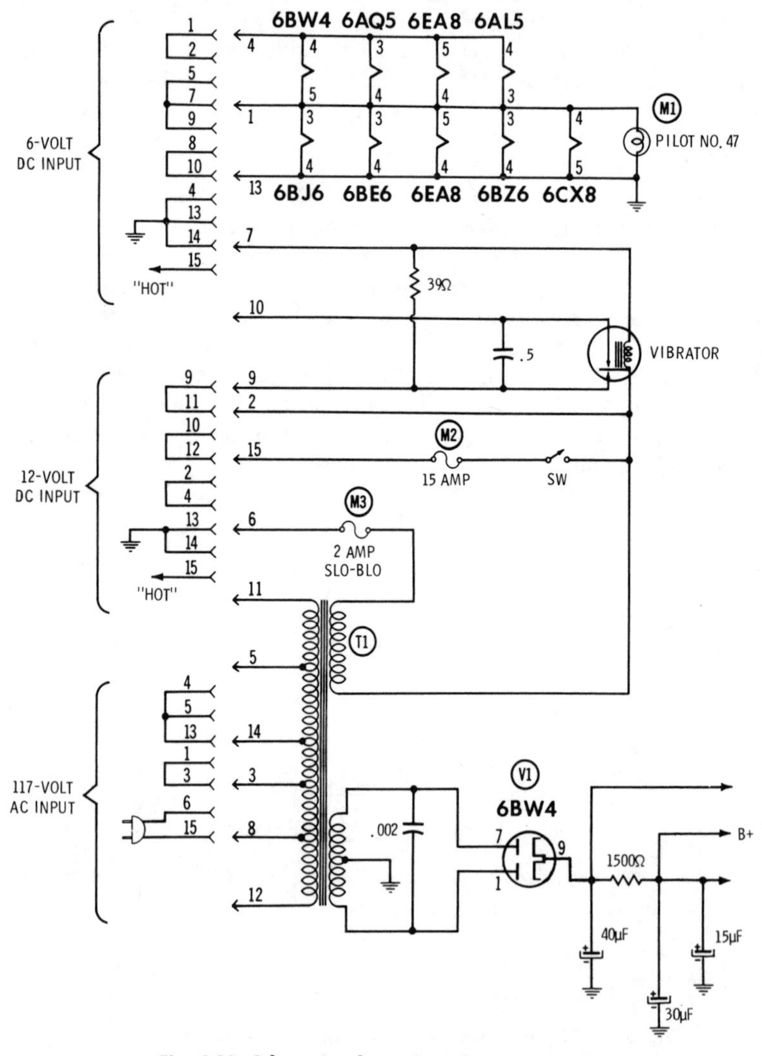

Fig. 3-30. Schematic of a universal power supply.

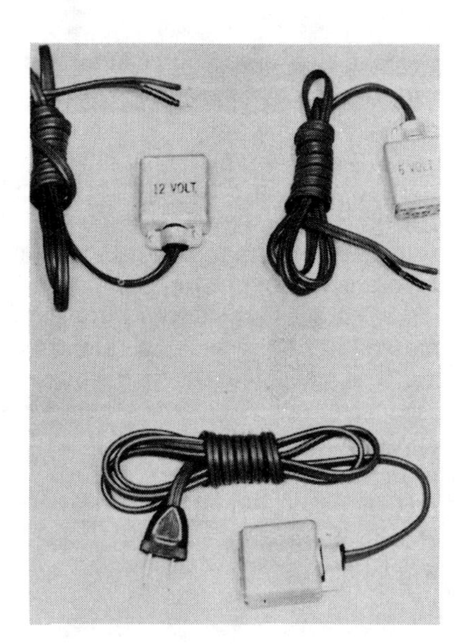

Fig. 3-31. Power cords used with the universal power supply in Fig. 3-30.

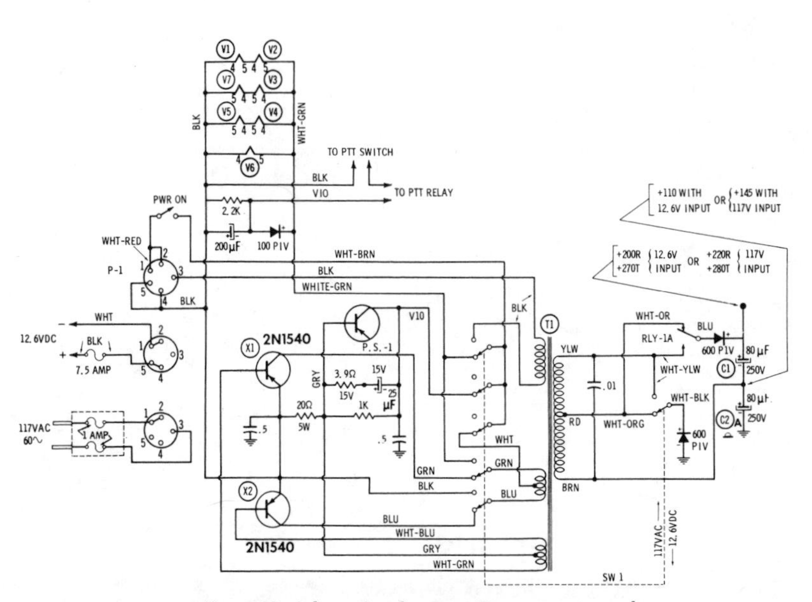

Fig. 3-32. Schematic of a transistor power supply.

tween the voltage source and this connector. Each plug is marked with its intended function so that, when connected to the transceiver, it will accommodate the desired input voltage.

Transistor Supply—To minimize the current requirements of the transceiver, some manufacturers use a transistor power supply that performs basically like the power supplies discussed previously—the principal difference being the manner in which the dc is converted before being stepped up by the transformer.

Fig. 3-32 shows one example of a transistor power supply. This particular supply will operate from either 12 volts dc or 117 volts ac. Basically, two matched power transistors (X1 and X2) operate as a push-pull oscillator when 12 volts dc are applied through the appropriate connector to P-1. The resultant alternating current is then applied to the transformer primary where it is stepped up, rectified, and filtered in the usual manner. When the supply is operated in the ac mode, power is applied directly to a separate primary winding.

4

Antenna Systems

The antenna is a vital link in the radio communications system; without some form of antenna, transmission and reception of signals would not be possible. All Citizens-band equipment has provisions for some type of antenna. It may be anything from a simple plug-in whip antenna to a complex rotary beam—depending on how the equipment is to be used and how much the user can afford to invest. The important thing is that a good antenna is a requisite for reliable communications over any substantial distance.

One of the biggest problems facing CB radio today is limited range. In many instances it can be directly attributed to improper antenna selection and installation; thus, proper choice of an antenna is a must for efficient operation. This is understandable when you consider the relatively weak signals handled by CB equipment. Antenna height is also important; however, this is not a matter of choice, but is limited by FCC regulations. Class-A stations are not nearly as restricted on this point as those in the Class-D service.

Also to be considered are the characteristics of radio waves at the frequencies used for CB operation. A better understanding of the problems involved will give you some idea of what to expect in the way of communicating range.

CHARACTERISTICS OF RADIO WAVES

All radio waves are electromagnetic in nature—that is, they are formed in much the same manner as the magnetic field that exists around a conductor through which current is flowing. However, radio waves actually radiate into free space somewhat like sound waves only much faster—in fact, approximately 186,000 miles a sec-

Fig. 4-1. Electromagnetic waves from the transmitter induce similar waves in the receiving antenna.

ond. Depending on the frequency and intensity of the antenna currents that produce them, radio waves can travel distances ranging up to several thousands and even millions of miles before their energy is dissipated.

Because of their electromagnetic properties, the waves leaving the antenna will induce currents in any conductor they contact (Fig. 4-1). Furthermore, these induced currents will exhibit the same characteristics as the currents originating at the transmitting station. From this you might assume that the transmission and reception of radio waves is fairly simple. However, there are a few other factors that complicate the picture somewhat. For example, an antenna of just any length will not respond efficiently to radio waves of a particular frequency. The frequency of a radio wave is determined by the number of cycles that occur each second, and the wavelength is the distance traveled during one cycle. This is illustrated in Fig. 4-2. Maximum current will be induced into the receiving antenna when its length has a specific mathematical relationship with the wavelength of the signal being received. The same relationship holds true with regard to the transmitted signal.

Wave Polarization

Wave polarization is determined by the angle at which radio waves leave the antenna. A perpendicular antenna will radiate ver-

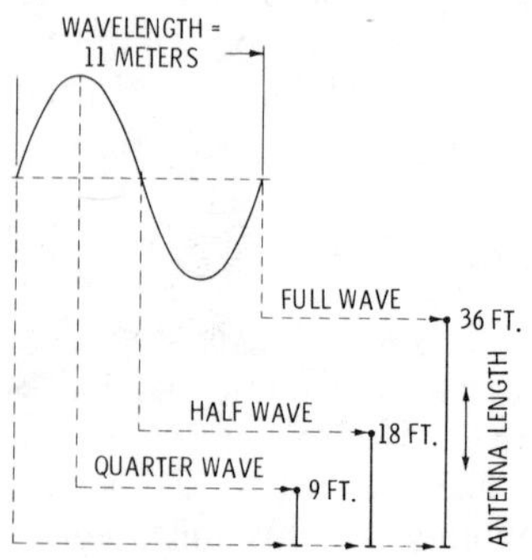

Fig. 4-2. Wavelength versus antenna length.

tically polarized waves; a horizontal antenna emits horizontally polarized waves. Vertical antennas are usually nondirectional—that is, they radiate and receive signals equally well in all directions. On the other hand, horizontal antennas have directional characteristics. The important thing to remember is that signal transfer will be maximum only when both stations use antennas with the same polarization.

Wave Propagation

One of the peculiar characteristics of radio waves is their behavior at different frequencies. For example, a transmitter operating in the 14-MHz range, even with relatively little power, is capable of communicating with stations thousands of miles away; others operating at uhf frequencies will have much more limited range.

There are two classifications for radio-wave travel—ground and sky. Ground waves travel along the surface of the earth, even to the extent of following its curvature to some degree. Sky waves can be explained more easily by referring to Fig. 4-3. If station 1, operating at approximately 14 MHz, attempts to contact station 2, ground wave A may never reach its intended destination. However, far above the surface of the earth are layers of ionized gases known as Kennelly-Heaviside layers. Under certain conditions, these layers will bend and reflect radio signals back to the earth. Thus, if conditions are right, the signals from station 1 will follow path B to arrive at station 2.

Since the ionization properties as well as the position of these layers vary, conditions may easily change the reflected signal path so that station 1 cannot communicate with station 2. Instead, path C may be established, and a station in this location could pick up the reflected signal which completely skipped this area when it followed path B.

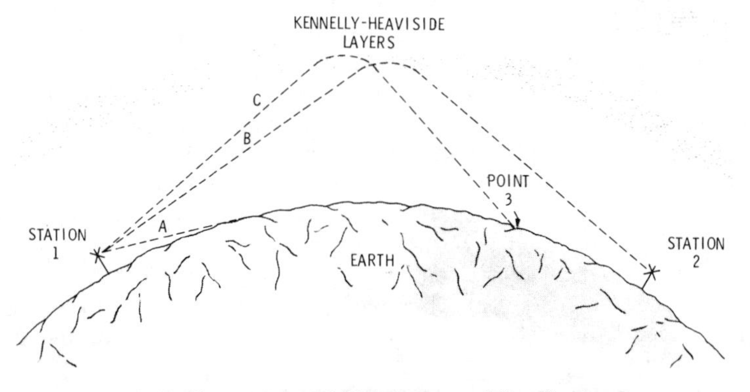

Fig. 4-3. Characteristics of high-frequency (hf) radio signals.

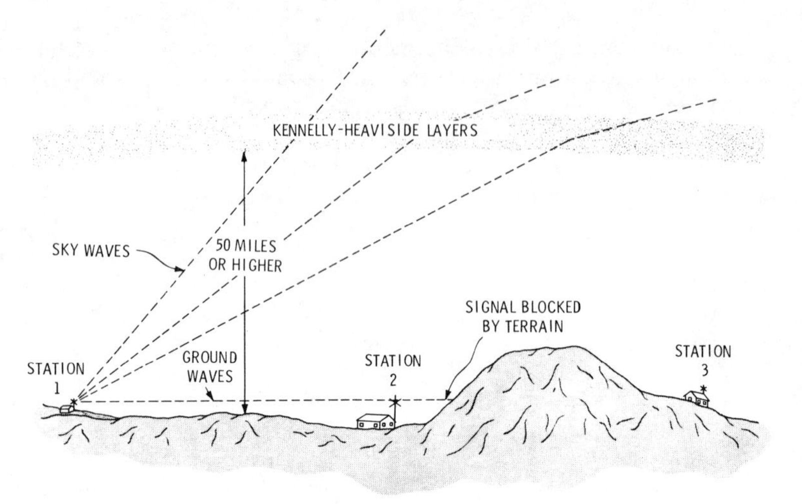

Fig. 4-4. Characteristics of ultrahigh frequency (uhf) signals.

Uhf signals, such as those used by Class-A stations, react quite differently. Wave travel is limited almost entirely to ground waves with line-of-sight characteristics—that is, they do not readily bend or curve, but follow a very straight path. With no obstructions between the transmitting and receiving antennas, communications can be achieved over surprising distances. However, intervening objects such as trees, buildings, terrain, etc., diffuse the signal considerably, and may even block it entirely. Fig. 4-4 better illustrates how uhf signals react. Ground waves from station 1 follow a line-of-sight path to station 2, providing good quality reception. However, the sky waves, unlike the lower-frequency signals in Fig. 4-3, are not bent back to earth by the ionized layers. Instead, they tend to travel straight through, or at best are deflected only slightly from their original course. If there were small obstructions such as trees, wooden buildings, etc., between stations 1 and 2 in Fig. 4-4, chances are the signal would still be received at station 2, but it would be considerably weaker—just as a beam of light is diffused when shining through a translucent substance. The amount of signal reduction depends on the number and type of intervening objects between the antennas. Obviously, the signal will be completely blocked as far as station 3 is concerned. At relatively close range, however, uhf signals are generally received regardless of obstructions.

Class-D frequencies are low enough to overcome some of the line-of-sight characteristics of uhf signals. Still, a direct path between antennas produces the best results. However, because of the relatively lower frequencies used here, there is less attenuation of the ground waves from intervening objects, thereby providing more reliable communications over a wider area. Even sky waves at these frequencies can produce "skip" reception of signals originating from thousands of miles away.

Antenna Gain

An antenna which is suitable for transmitting is equally effective as the receiving element. Some are designed with directional characteristics (Fig. 4-5); others are omnidirectional (respond to signals from any direction). The best way to determine the operational characteristics of an antenna is from its radiation pattern. This pattern is often provided on the manufacturers specification sheet.

Fig. 4-5. The VGR-27 Class-D directional ground-plane antenna.

Courtesy Mosley Electronics, Inc.

An antenna with directional characteristics usually provides a certain amount of gain. In fact, through careful design, an omnidirectional antenna can also be made to provide gain. Although we speak of an antenna as having a certain gain, it should not be misinterpreted as meaning that it can actually amplify radio signals like a transistor or a vacuum tube. No antenna can provide gain in that sense. However, through proper design, an antenna can be made to concentrate its radiated energy in such a way that it will appear to have been produced by a much stronger source. You cannot get something for nothing though, so increasing the field strength in one direction is only made possible by reducing it elsewhere.

Before any gain can be realized with an omnidirectional antenna, it must be able to concentrate a portion of its radiated rf energy in such a manner as to add to, or reinforce, the existing energy already following the desired path. This is accomplished by lowering the angle of radiation so as to divert that energy which would normally be lost as sky waves.

The gain of an antenna is determined by comparing its performance with that of a standard antenna, and expressing this figure as a ratio of the power levels required to produce equivalent field strengths. The gain, then, in decibels equals ten times the log of this power ratio. An antenna that provides a 3-dB gain, for example, will effectively double transmitter power. In other words, a trans-

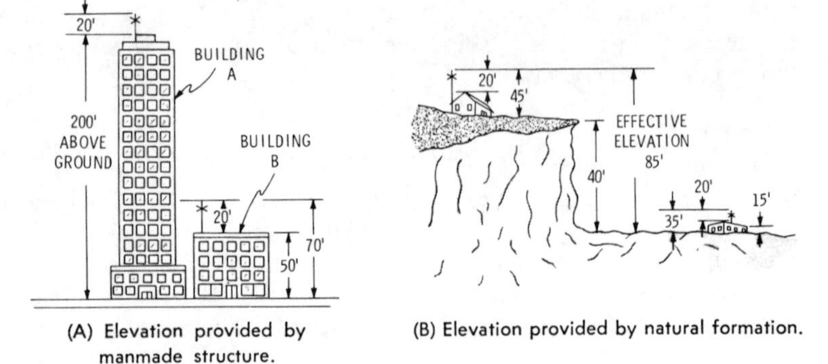

(A) Elevation provided by manmade structure.

(B) Elevation provided by natural formation.

Fig. 4-6. Examples of how antenna height compares with effective elevation.

mitter with 3-watts-output feeding such an antenna, will "look" to the receiving station like a transmitter radiating 6 watts.

HEIGHT LIMITATIONS

Antenna height regulations adopted for the Citizens Radio Service are given in Section 95.37 of the FCC regulations (see Appendix). Class-A stations are permitted to use relatively high antenna structures, but are limited when such installations may become a hazard to air navigation. The antenna height of Class-C and -D stations operating as fixed stations, however, cannot exceed 20 feet above the

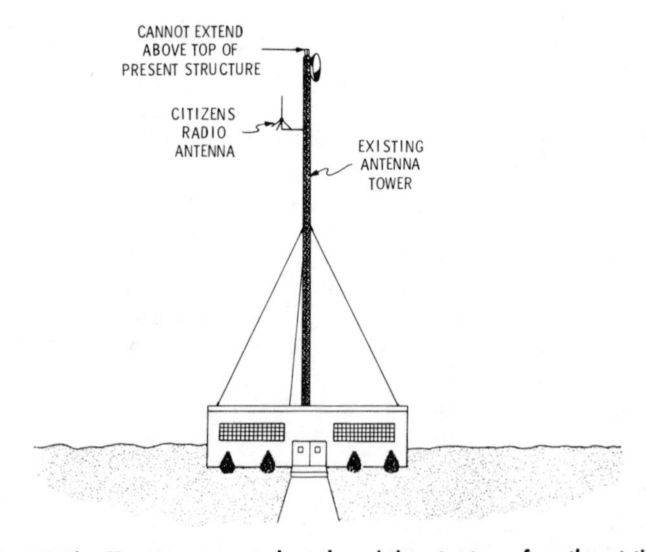

Fig. 4-7. The CB antenna mounted on the existing structure of another station.

man-made or natural structure on which it is mounted. A further limitation to this rule stipulates that when a CB antenna is mounted on an existing antenna structure of another station (such as a tower), it cannot extend above it by *any* amount.

Fig. 4-6 shows how the 20-foot restriction applies. The buildings in Fig. 4-6A are of unequal heights. Despite the difference, the antenna structure must not extend over 20 feet above its mounting. Both antennas have the maximum allowable height, yet the antenna on building A has an effective elevation of 220 feet as compared with 70 feet for the one on building B. Fig. 4-6B shows another example. The antenna in this case is mounted above a house situated on a hill, providing a much higher effective elevation than the station at the lower level using an installation of similar height. Fig. 4-7 shows the one exception of the 20-foot rule, where the antenna is mounted on the existing antenna structure of another station. It can be mounted at any point on this tower, below the existing antenna.

PHYSICAL ASPECTS

The type of CB operation you plan will undoubtedly influence your selection of an antenna. For example, if you intend to communicate only over an extremely short distance, perhaps with a neighbor who lives one or two blocks down the street or in the same building, you would hardly benefit from buying a beam antenna when a simple whip antenna will suffice. Moreover, the size of the antenna may have a bearing on the type of installation you will want.

Antennas are most effective when cut to a size equal to one wavelength or a submultiple (such as a half or quarter wavelength) of the signal frequency. Since the Class-D range covers a number of frequencies, the most practical solution is to use a compromise length, cut for the center of the band. A wavelength in the 27-MHz Class-D range would be about 11 meters, or over 36 feet long. Obviously, an antenna of such length would be impractical for most installations. Thus, most 27-MHz antennas are cut to either a half or quarter wavelength. Even so, an 11-meter half-wave antenna is approximately 18 feet long, and a quarter-wave around 9 feet.

Loading Coils

The physical length of a vertical antenna can be made shorter by use of a loading coil, a device which effectively increases the electrical length of the antenna. Such an arrangement provides a solution where excessive length is a problem. Loading coils are used primarily with mobile antennas.

The loading coil may be located at the base, center, or top of the radiating element. Fig. 4-8 shows examples of typical base-loaded

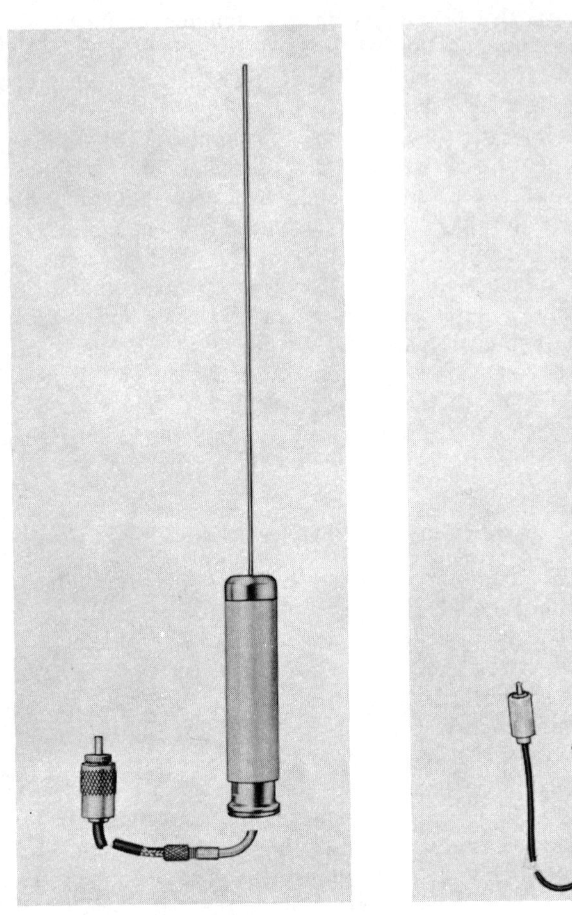

(A) A mobile whip (M-67) using a base loading coil.

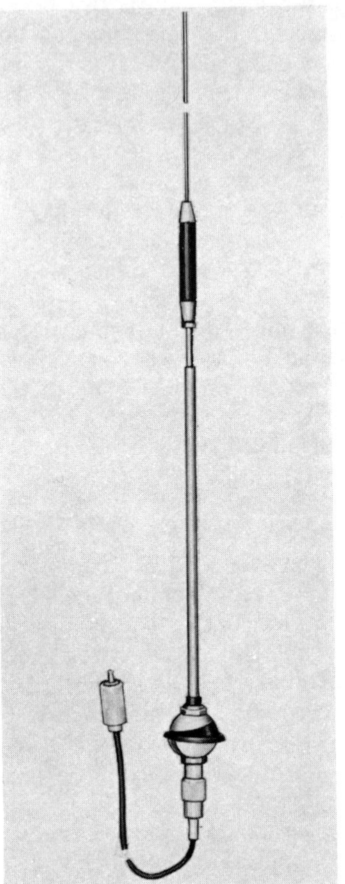

(B) A center-loaded mobile whip (M-49).

Courtesy The Antenna Specialists Co.

Fig. 4-8. Examples of mobile whip antennas using loading coils.

and center-loaded mobile whip antennas. A fourth type of loading, known as continuous, is accomplished by spiral-winding the inductor along the entire length of the radiating element. An example of this type of antenna is shown in Fig. 4-9.

ANTENNA SELECTION

A variety of antennas is used for CB communications, most of which fall into one of four basic categories—whip, coaxial, ground

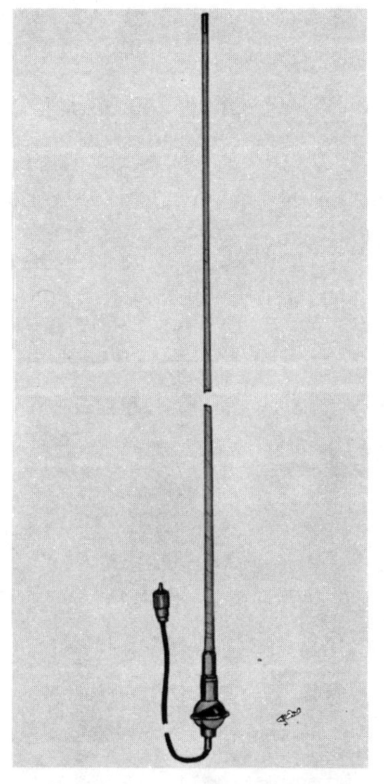

Fig. 4-9. A continuously loaded mobile whip antenna (M-52).

Courtesy The Antenna Specialists Co.

plane, and beam. An antenna may be included with the CB gear when purchased, or it may be available as an accessory. Those included will usually be the simple plug-in type of fixed or adjustable length, and may or may not include a loading coil. Furthermore, some antennas are removable while others are actually a physical part of the transceiver itself, as in the case of some of the smaller transistorized units. These antennas are designed primarily for short-distance communications, and while they work quite well in the open, they may not provide satisfactory results for the type of coverage you desire.

Proper antenna length is important if the available transmitter power is to be used to the utmost advantage. If communications between two fixed stations are to be accomplished over a considerable distance, beam antennas can be used to great advantage. Not only do they provide a signal gain, but they generally discriminate against unwanted signals from other directions. When communications are desired between two stations only, the beams should be directed toward each other and secured in this position. However, when communication with other stations is necessary, a rotator (Fig. 4-10) may be installed, permitting the antenna to be aimed in any direction.

Wave polarization must also be considered, since communications between stations using unlike polarization will be poor. When a fixed station is to communicate with mobile units, it should be equipped with an antenna that will provide equal response on all sides. Practically all mobile antennas are designed to radiate a vertically polarized wave; therefore, the antenna at the base station must also be vertically polarized. A vertical beam antenna mounted on a rotator will do the job, but the ground plane or the coaxial antenna is used most often in this application.

Courtesy Hy-Gain Electronics Corp.

Fig. 4-10. Example of heavy-duty antenna rotator designed for use with large beams and stacked arrays.

The ground plane has a low angle of radiation which confines most of the signal to ground-wave paths and allows little energy to escape into the ionosphere. The popular coaxial antenna (sometimes referred to as a "thunderstick") also exhibits good omnidirectional characteristics and is used to advantage in fixed-station installations where space is a problem. In addition, the coaxial antenna is very easy to install. Fig. 4-11 illustrates several examples of CB base station antennas that are currently being offered.

ANTENNA INSTALLATION

In many cases the simple vertical type of antenna will serve very well in the intended operation, and no special installation will be required. However, if an outside antenna is to be used, a certain amount of planning and physical labor will be required. This may be performed either by a service technician or by the licensee himself.

Fixed Station

Since Class-A stations are not too restricted in the way of antenna height, they often employ rather complex structures which require special equipment to erect. An example of a typical base-station installation suitable for Class-A operation is shown in Fig. 4-12. Obviously, an installation such as this must be properly guyed and anchored to withstand high winds. This one is guyed from three angles, as shown at the upper left in the illustration. Insulators are inserted into the guy wires about 20 feet above the anchors, and turnbuckles are provided to take up any slack. A coaxial cable of the proper impedance is placed along the inside of the tower and fastened about every six feet before being connected to the antenna mounted at the top. Here a vertical (nondirectional) antenna is used; other types may be employed.

Outside antennas designed for Class-D operation are generally no more difficult to install than the average television antenna of equal height. It is naturally desirable that the antenna be erected as high as possible, especially in lower-level areas where full advantage should be taken of the 20-foot height permitted. However, if the station is situated on a sufficiently high elevation to begin with, you may obtain satisfactory results with less than the full 20 feet—perhaps eliminating the need for a tall mast and guy wires. Fig. 4-13 shows two typical Class-D ground plane antenna installations. A heavy-duty ground-plane unit is used in Fig. 4-13A and Fig. 4-13B shows a collinear type—a version of the ground plane in which the radials are placed at right angles with the vertical center element.

To further increase operating efficiency, a stacked beam or ground-plane antenna can be used. Here, two separate units are placed one above or beside the other on a single mast, in the same manner as many television receiving antennas. Fig. 4-14 shows a pair of horizontally stacked five-element vertical beams used in Class-D operation. This same array could also be stacked vertically.

The ground plane can likewise be stacked, but only vertically. A dual arrangement of this type generally lowers the angle of radiation even further than the single unit.

Stacked arrays for Class-D operation can be purchased as a unit —that is, matching antennas complete with hardware and assembly instructions, or as separate units. So if a single-bay beam, for example, has already been installed and you now wish to add another bay, it should present no problem.

Although it is desirable for the antenna to be installed as high as possible within the legal limit, there is another factor that must be considered—the length of the coaxial cable between the transmitter and antenna. This cable consists of a center conductor en-

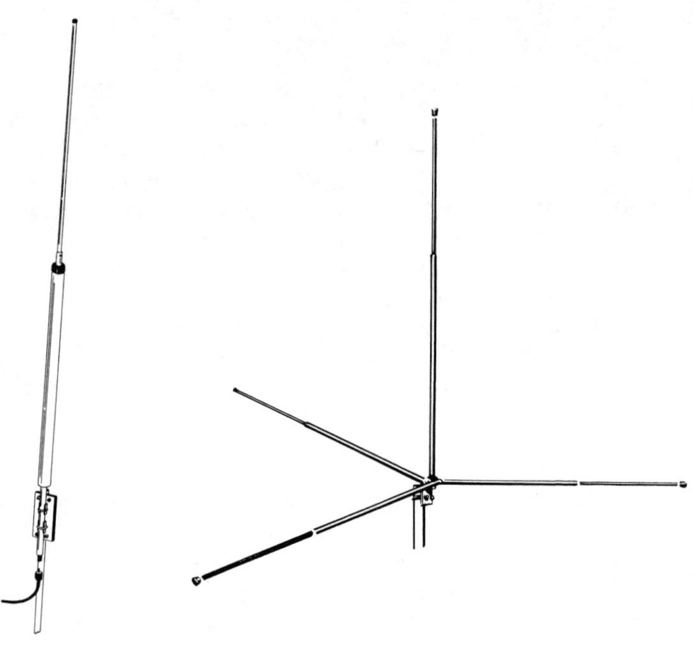

Courtesy Mosley Electronics, Inc.
(A) Coaxial antenna. Provides unity gain and omnidirectional pattern.

Courtesy Hy-Gain Electronics Corp.
(B) Quarter-wave ground plane. Characterized by omnidirectional, unity gain pattern.

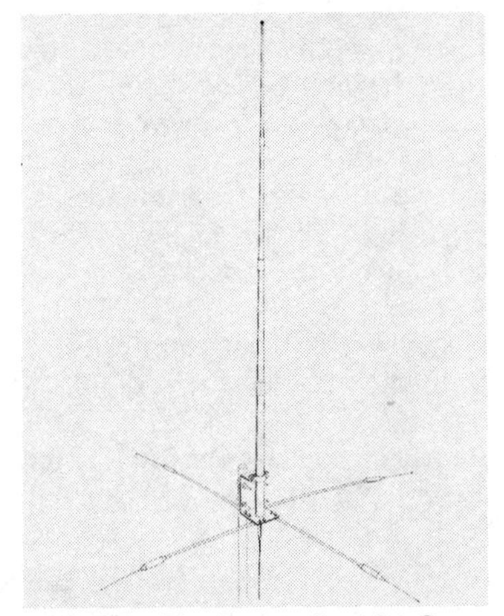

Courtesy The Antenna Specialist Co.
(C) The "Mag X" provides 4 dB of omnidirectional gain.

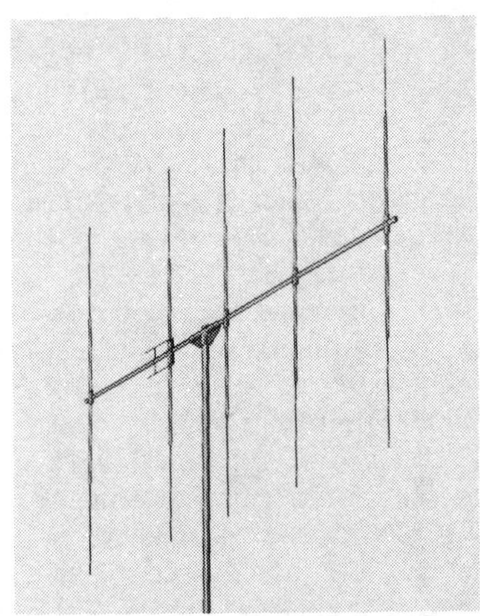

Courtesy Hy-Gain Electronics Corp.
(D) The "Long John" wide-spaced, 5-element beam is a directional antenna that provides 12.3 dB power gain.

Fig. 4-11. Examples of Class-D base

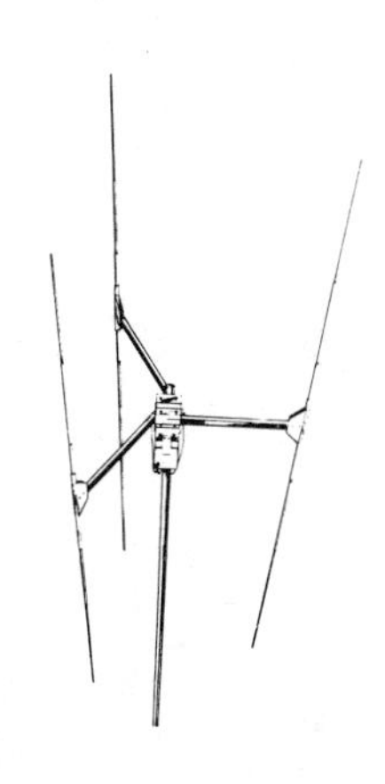

Courtesy The Antenna Specialist Co.
(E) Model MR-119 "Super Scanner" provides choice of either directional or omnidirectional pattern and up to 7.75 dB of gain.

Courtesy The Antenna Specialist Co.
(F) "Boss 404" features yagi design, high gain, and selection of either vertical or horizontal polarization.

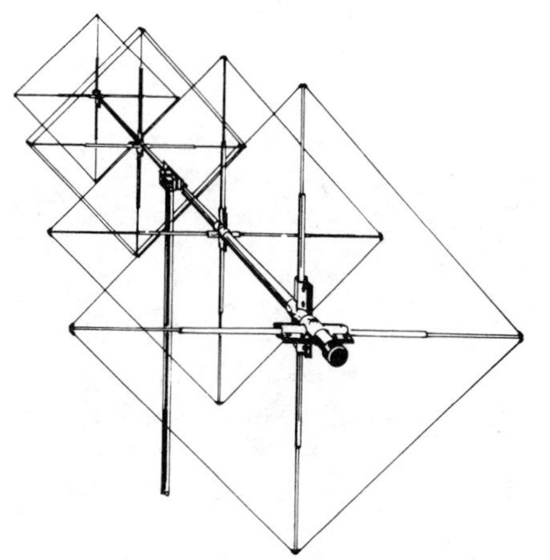

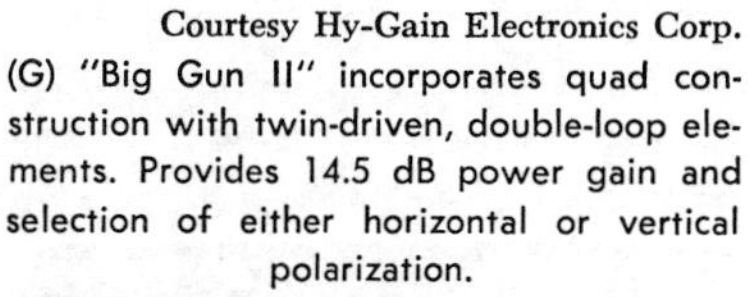

Courtesy Hy-Gain Electronics Corp.
(G) "Big Gun II" incorporates quad construction with twin-driven, double-loop elements. Provides 14.5 dB power gain and selection of either horizontal or vertical polarization.

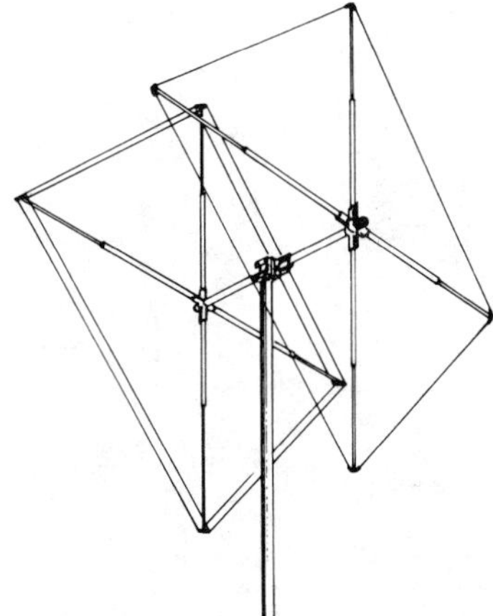

Courtesy Hy-Gain Electronics Corp.
(H) The "Eliminator II" quad provides selection of antenna polarization and provides a power gain of 9 dB.

station antennas currently being offered.

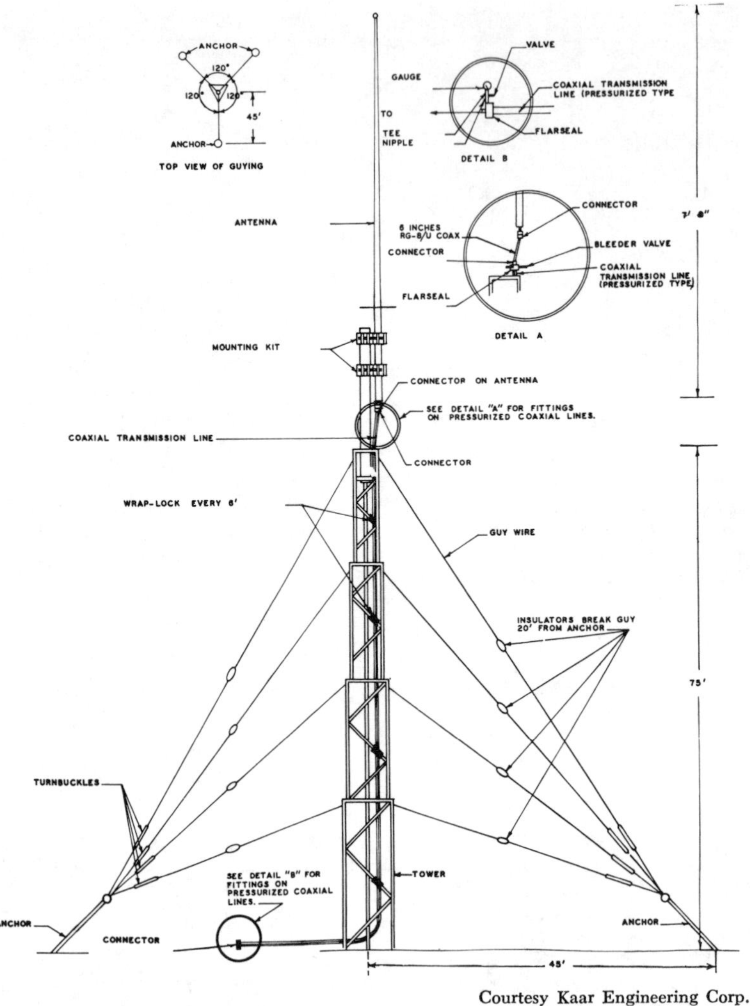

Courtesy Kaar Engineering Corp.

Fig. 4-12. Typical antenna installation for a Class-A base station.

cased in polyethylene or similar type of dielectric material. Around this is a braided shield, which in turn is covered by a vinyl jacket. The outside cover insulates the entire cable (Fig. 4-15). The braided shield serves not only as the other conductor, but also as a shield against interference, and keeps the rf energy passing through the center conductor from being radiated before reaching the antenna.

The important factor here is the rf loss of the cable. The manufacturer's specifications usually state the loss of a particular cable

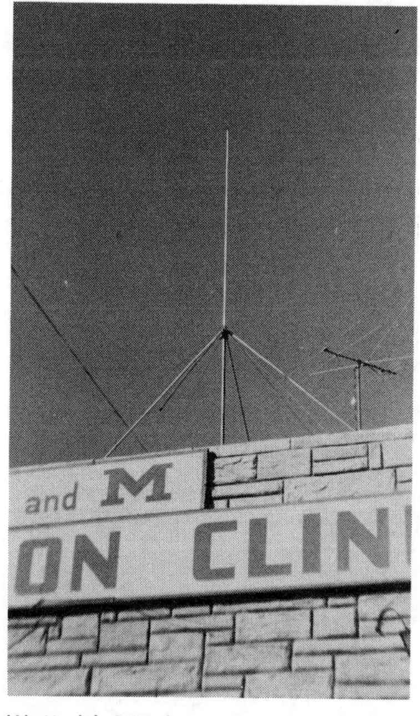

(A) Model GP-1 heavy-duty ground plane.

(B) Model CLR collinear ground plane.

Courtesy Hy-Gain Electronics Corp.

Fig. 4-13. Typical ground plane antenna installations.

as so many dB of attenuation per 100 feet at a given frequency. The higher the frequency, the greater the rf loss; also, the longer the cable, the higher the loss.

If you plan to install your own antenna, be sure to check the equipment manual for the type of coax to be used. Maximum power is transferred between the transceiver and antenna when their impedances match. Using a type of coax other than that recommended can result in an impedance mismatch and subsequent loss of efficiency.

Since a longer transmission line introduces more signal attenuation, it should be kept as short as possible. However, if running a cable another 50 to 100 feet would increase the effective elevation of the antenna another 25 feet, the additional range—which may be as much as 50 percent—would overshadow the additional rf cable loss.

Even though a certain amount of rf loss is incurred in the transmission line, it can often be offset to some degree by using a more efficient antenna. Another solution is to reduce the length of the transmission line as much as possible by mounting the radio equip-

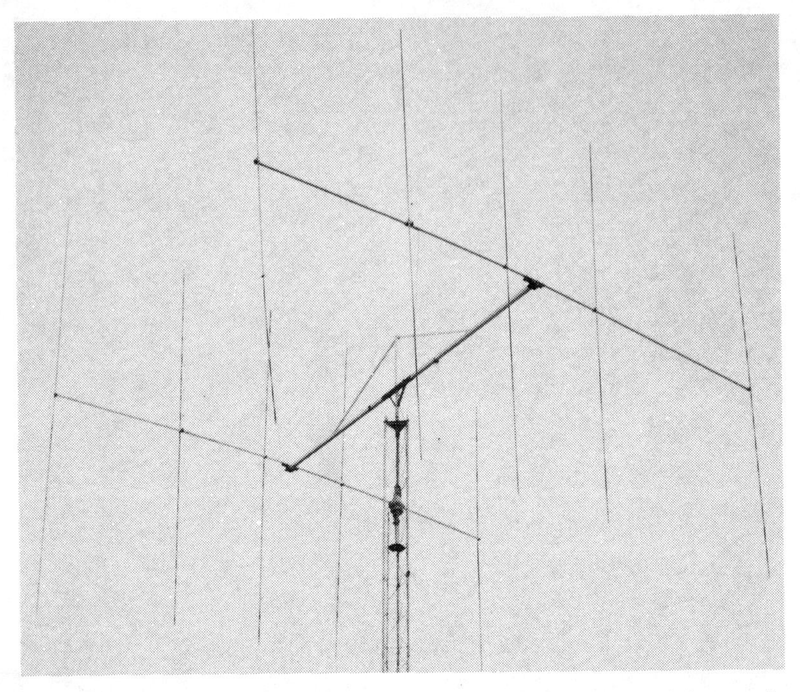

Fig. 4-14. Example of CB antenna installation incorporating an array of horizontally stacked, five-element vertical beams.

ment at the base or near the top of the antenna support as shown in Fig. 4-16. It can then be operated from some remote location. Obviously, steps will have to be taken to ensure a waterproof installation; otherwise the transceiver could be damaged.

The term "remote control," when applied to the operation of a Citizens radio station, means control of that station transmitter from any place other than the location of the transmitting equipment itself. Remote control of this type is allowed only to Class-A base or fixed stations; however, Classes C and D can use direct mechanical control or even direct electrical control by means of wired connections to the transmitter, as long as it is located on the same premises. For example, if another building on the same property will provide more height, the transceiver can be installed there and controlled (in the manner just described) at any other point on the same property. The transceiver illustrated in Fig. 4-16 is controlled from inside the building by means of a multiconductor cable. If the transceiver

Fig. 4-15. Construction of a coaxial transmission line.

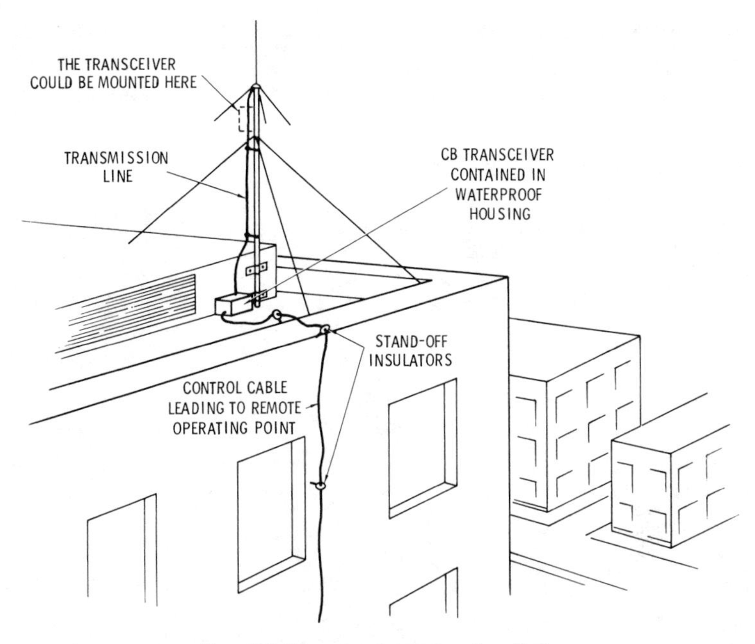

Fig. 4-16. A remote transceiver installation.

is mounted near the top of the mast, additional guying will be required.

Mobile

Mobile operation does not necessarily mean a station installed in a vehicle. It is defined as a station used while in motion or during halts at unspecified points. All units in Classes C and D are classified as mobile stations, regardless of whether they are permanently located or installed in a vehicle. However, since fixed stations have already been discussed, let's concentrate for the moment on vehicle installations.

Vertical whip antennas are most commonly employed in auto installations because they require little space and are easily installed. Many automobile whips employ some type of loading coil to reduce their physical length; otherwise, they would be rather awkward.

This type of antenna can be mounted in a variety of places, as shown in Fig. 4-17. The location will depend on the type of vehicle. The cowl mounting shown in Fig. 4-17A is no more trouble to install than a regular auto-radio antenna. A roof mount is illustrated in Fig. 4-17B; although it affords maximum height, it is prone to damage from striking trees and other objects while the vehicle is in motion. This mounting location is most desirable, however, and can

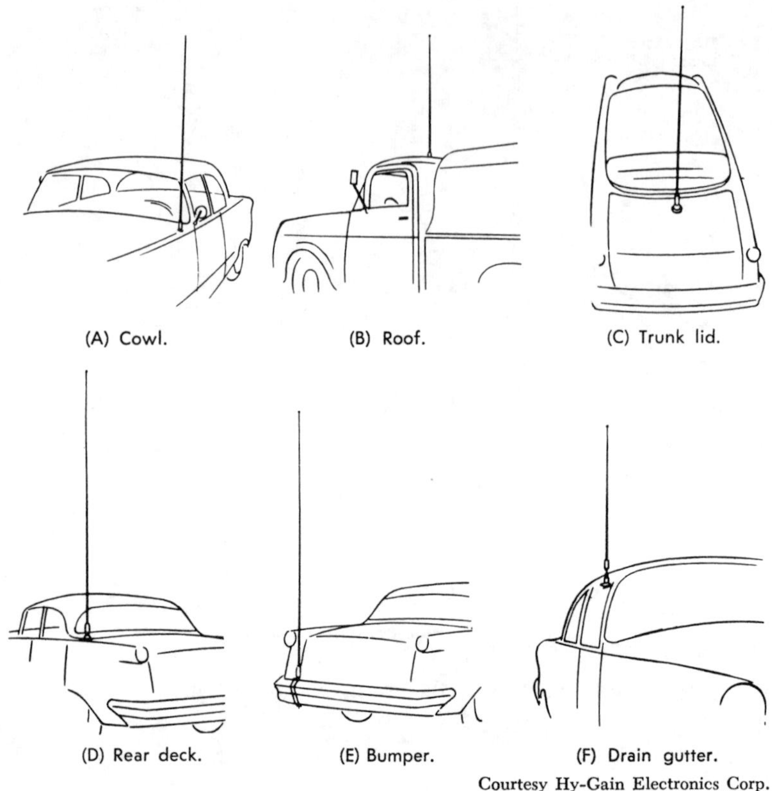

(A) Cowl. (B) Roof. (C) Trunk lid.

(D) Rear deck. (E) Bumper. (F) Drain gutter.

Courtesy Hy-Gain Electronics Corp.

Fig. 4-17. Typical mobile whip mounting locations.

be used to good advantage when a shortened (loaded) antenna is to be employed. The whip antenna can also be installed on the trunk deck of some cars (Fig. 4-17C). This installation, like the one in Fig. 4-17D, provides a compromise in overall antenna height between the roof mount (Fig. 4-17B) and the bumper mount (Fig. 4-17E).

The bumper mount is illustrated in Fig. 4-17E. The whip and mounting are connected to a clamp or strap arrangement that fastens around the bumper. This is one of the simplest whip installations and generally requires no hole drilling. The mounting arrangements in Figs. 4-17D and E are two of the most popular in use today. Fig. 4-17F shows a quick and easy whip installation for close-range operations. Here, the antenna is clamped to the gutter at the top of the door.

The mounting location has considerable effect on the performance of the antenna because the metal car body will influence the radia-

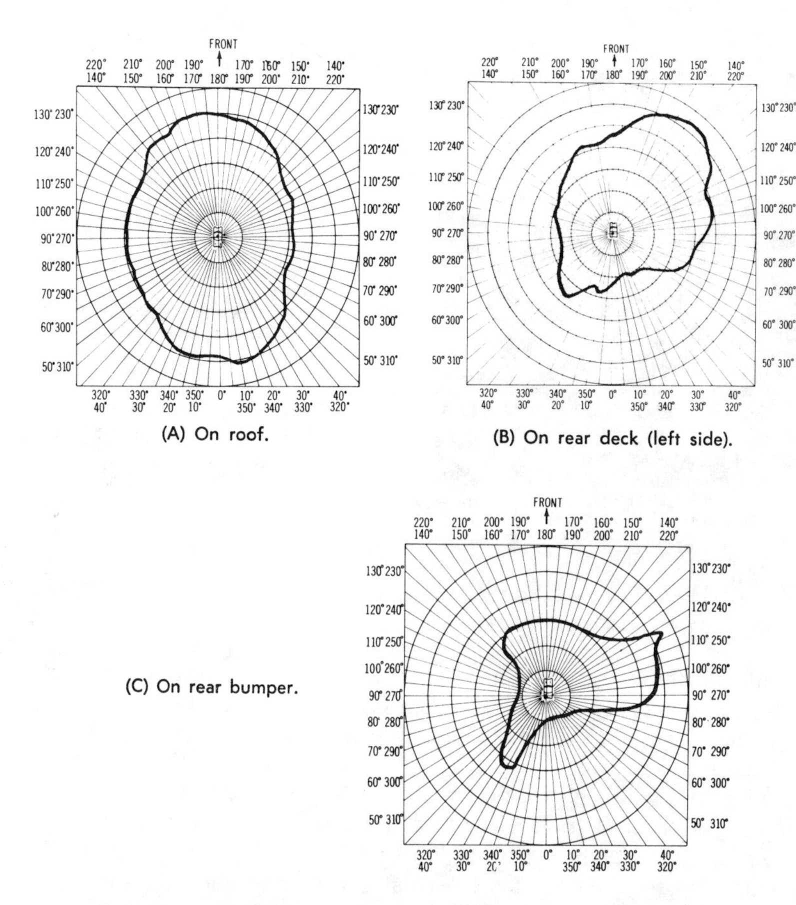

(A) On roof.

(B) On rear deck (left side).

(C) On rear bumper.

Fig. 4-18. Horizontal radiation patterns when whip antenna is mounted at various locations on the vehicle.

tion and reception pattern as illustrated in Fig. 4-18. In all three cases a quarter-wave vertical whip antenna is employed; the only difference is the mounting location. Thus, it is obvious that the most desirable mounting location is the roof of the vehicle. However, this is not always possible. Much will depend on the type of vehicle, size of the antenna, etc.

Regardless of where the mounting is located, the antenna cable should be kept as short as possible and routed away from the engine, gauges, wiring, and other sources of interference. Even though shielded, the cable may pick up some interference due to poor connections, etc.

A standard automobile broadcast antenna should never be used with a CB transceiver. It will introduce considerable rf loss and

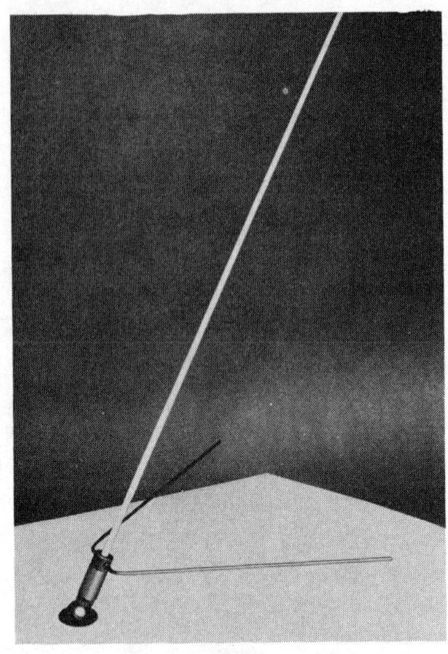

Courtesy Karr Engineering Corp.

Fig. 4-19. The "Sea Sprite" designed for Class-D marine installations.

result in inferior performance. There are, however, several antennas designed especially to serve the dual function of standard a-m broadcast and CB radio antenna. These antennas are peaked for maximum CB performance but operate equally well for standard broadcast reception. A dividing harness permits either separate or simultaneous operation of the CB transceiver and standard broadcast receiver.

A Class-D antenna designed especially for boat installations is pictured in Fig. 4-19. This one is five feet long and employs base loading. It is constructed of *Fiberglas* to resist the corrosion usually experienced in salt-water areas. This antenna can be obtained with either a permanent or temporary mounting. The former is a swivel-type mounting device similar to that used on many standard auto-radio antennas. The temporary mounting has rubber suction cups which allow it to be attached to any smooth surface—a great convenience if the radio equipment is to be used in more than one location.

COMMUNICATING RANGE

There are no set figures to indicate just how far Citizens-band transmissions will carry. Even for a given input power, the communicating range will vary greatly.

Range is determined by a number of factors, one of which is the type of CB equipment being used. A Class-A unit operating as a fixed station with a power input of 5 watts and with the proper antenna is capable of line-of-sight communications of up to 10 miles or more with similarly equipped stations—even though uhf frequencies are used and wave travel consists almost entirely of line-of-sight ground waves. Under adverse conditions, transmissions from the same station may be effective up to only a mile or so if operating from a moving vehicle in an area where buildings and other obstructions are present. These can attenuate the uhf signal severely, although short-distance transmissions are usually effective despite such obstructions. Also, relocating the antenna even a slight distance away may improve uhf reception considerably.

Class-A equipment operating with the maximum 60 watts input, of course, will provide somewhat better range. However, the communicating range does not increase proportionately with power, especially at uhf frequencies. Many Class-A units are designed to operate at less than the maximum input power allowed, because full power operation will not produce a substantial increase in range. Most full-powered Class-A equipment is capable of providing communications ranging from 10 to 15 miles or more.

With Class-D equipment, communications between fixed stations separated by 20 miles or more is not uncommon. Occasionally, atmospheric conditions are such that sky waves may be reflected back to earth thousands of miles away.

The range of Class-D base-to-mobile communications is normally 5 to 10 miles. It may be less than one mile where there are obstructions to block the ground waves, or more than 10 miles when the mobile unit operates from an elevated area.

The range of mobile-to-mobile communications using Class-D equipment also varies greatly—generally about 1 to 5 miles on land, and around 10 to 12 miles across water. Again, these figures vary considerably, depending on conditions.

Basically, seven factors determine the communicating range—type of equipment employed, transmitting power, operating frequency, surrounding terrain, amount of interference, type of antenna used, and antenna elevation. The latter does not necessarily mean the elevation of the antenna above ground, but rather its relative height with respect to the receiving antenna.

Does your intended equipment have the full input power permitted? And if so, is the efficiency of the transmitter such that a relatively high power output is obtained? The percentage of modulation also has some bearing on the output power and subsequent range. For example, a unit capable of 90 percent modulation will have the edge over one providing only 70 percent.

We have already discussed the effects of terrain and transmitting power on range, and have seen how signals react at various frequencies; but nothing has been mentioned about interference. One type, adjacent-channel interference, is troublesome in Class-D operation, where carrier frequencies are only 10 kHz apart. Adjacent-channel interference can be eliminated in most instances by using a receiver with good selectivity. However, other forms of interference, over which you have no control, may also be present. Examples are stations operating on the same channel, atmospheric disturbances (lightning, etc., to which a-m receivers are particularly susceptible), and interference from scientific, industrial, and medical equipment. Ignition and many other types of noise-pulse interference are more or less suppressed by the receiver's noise-limiter circuit (if one is used), but even it may not provide sufficient rejection under severe noise conditions.

The design of the antenna and its effective elevation also help to determine the communicating range. The type of antenna employed is a matter of personal choice; its height is limited by the FCC. For a more comprehensive discussion of CB antenna systems, refer to *CB Radio Antennas,* published by Howard W. Sams & Co., Inc.

5

Station Installation

One of the primary considerations for a base-station installation is the location of the antenna—particularly if maximum range is desired. Thus, it would seem advisable to place the antenna as high as possible; however, a long cable run to the transceiver may result in too much signal loss. The transceiver should also be situated relatively close to the power source, but since antenna elevation is most important, it may be more practical to place the unit closer to the antenna and use an extension cord of suitable current-carrying capacity to supply the required source voltage.

BASE STATIONS

Fig. 5-1 shows a typical Class-D fixed-station arrangement. In any installation of this type, the transceiver should be located as close as possible to the point where the antenna cable enters the room. After the antenna has been properly installed in an advantageous location, the coaxial transmission line should be routed directly to the transceiver. Usually it can be run under a window or through an access hole in the wall. In the latter case, some type of vitreous or ceramic tubular feedthrough insulator should be used (Fig. 5-2) to protect the cable, and the excess space around the insulator should be filled with a sealer. Running the lead between the window and the sill is not too desirable as a permanent installation. If the transmission line is longer than necessary, it should be cut to length and the transceiver end terminated with the proper antenna connector as shown in Fig. 5-3. Notice that the center conductor of the cable passes through the center terminal of the plug, and that the cable shield connects to the outer housing. Care must be taken when sol-

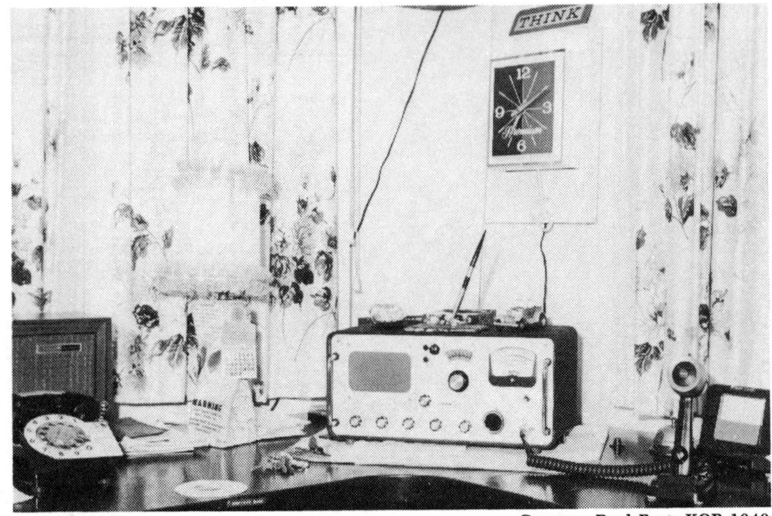

Courtesy Paul Bost, KQP 1040

Fig. 5-1. Typical Class-D base-station setup.

dering the cable braid and plug connections, since excessive heat will melt the polystyrene insulation between the center conductor and the shield. The power line, which may have to be extended to reach an outlet, should be routed neatly along the baseboard or under the edge of the carpet. The equipment itself can be placed on almost any level surface; however, it should be situated where it can receive

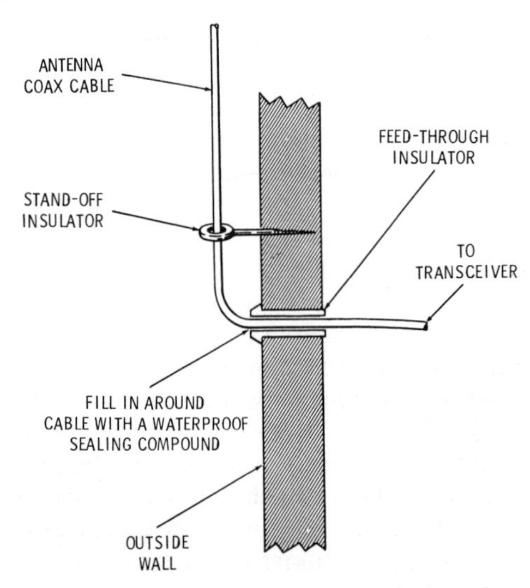

Fig. 5-2. Method of feeding the antenna cable through a wall.

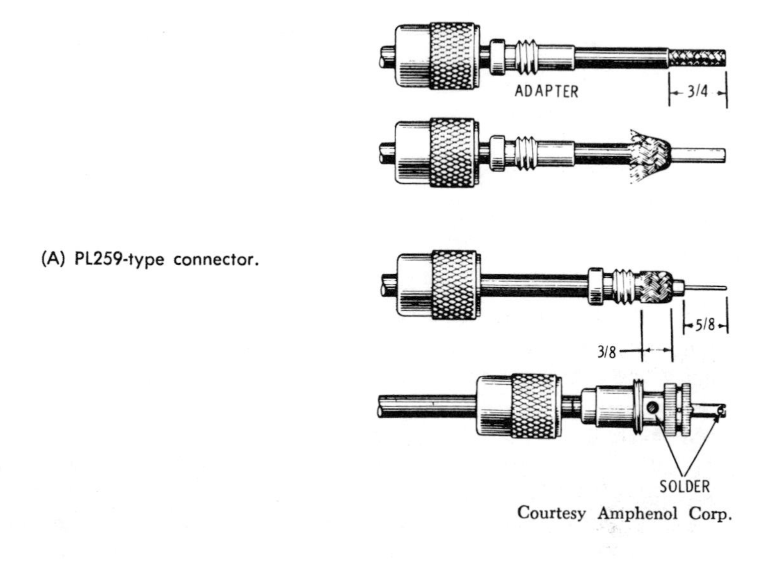

(A) PL259-type connector.

Courtesy Amphenol Corp.

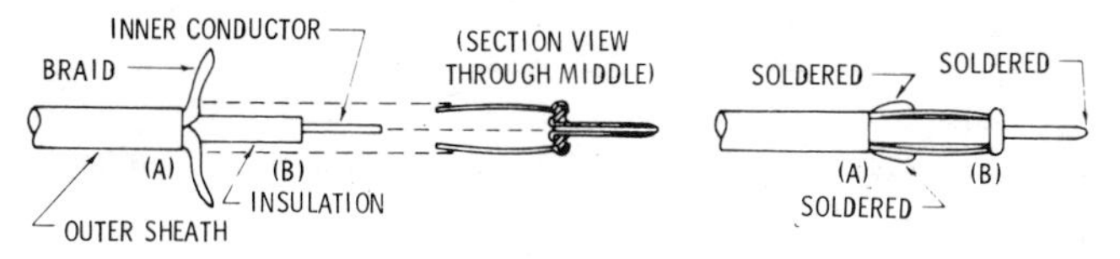

(B) Motorola-type connector.

Fig. 5-3. Two methods of fastening the antenna connector to the cable.

adequate ventilation to prevent overheating. Heat is seldom a problem with solid-state equipment, however.

Transceivers also require a good ground connection, which can be provided by driving a standard 5-foot copper ground rod into the earth and running a wire (not less than 12 gauge) between it and the proper terminal on the radio equipment. In some cases an existing water pipe or faucet can serve as the ground, with the wire firmly clamped to it. The antenna mast should also be grounded as protection against lightning.

A device specifically designed for lightning protection is shown in Fig. 5-4. When placed in series with the standard PL-259/SO-239 antenna cable connectors, and provided with the proper ground connection, this device provides protection for both the operator and the radio equipment from lightning and static electrical charges.

Another useful accessory for fixed-stations is the "Base Control" in Fig. 5-5. This unit combines numerous accessories and functions into one package to permit overall control from one location. Included are SWR bridge, modulation-percentage meter, power meter,

Fig. 5-4. A lightning protection device to safeguard both operator and equipment.

Courtesy Hy-Gain Electronics Corp.

phone patch, receiver preamp, built-in speaker, speech processor, antenna switch, and more. A variety of accessories are available to enhance your station. These are only two examples.

After all the required connections have been made, refer to the procedure outlined by the manufacturer for the initial setup and operation of the particular equipment being used. Generally, an antenna loading or matching adjustment of some type must be made prior to going on the air. Adjustments of this type are covered in Chapter 7.

Most CB equipment is of the transceiver type where the transmitter, receiver, and power supply are incorporated as a single unit. There are also base stations that consist of separate units which are interconnected to permit single-button operation. Practically all CB equipment (regardless of type) is designed to use the same antenna for both transmitting and receiving.

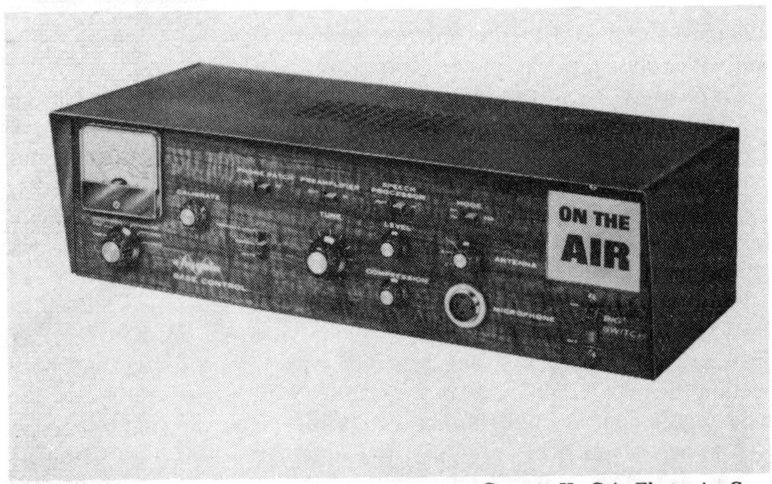

Courtesy Hy-Gain Electronics Corp.

Fig. 5-5. The "Base Control" incorporates numerous station accessories and permits overall station control from a single panel.

MOBILE INSTALLATIONS

Unlike the average fixed station, mobile installations will present individual problems because of the variety of vehicles in which the equipment may be installed. Although it is virtually impossible to give the exact procedure for each type of installation, an understanding of some of the major considerations will serve the purpose in most cases.

Installing CB equipment in an automobile or other vehicle is not difficult. Even a layman will have little trouble when guided by a few simple hints. Furthermore, only the basic complement of tools is needed—assorted screwdrivers (standard and Phillips head), socket wrenches, pliers, ¼-inch electric drill, a small soldering iron, and some rosin-core solder. A tapered reamer or a chassis punch can also be used to advantage for making oversized holes.

Mounting the Transceiver

The transceiver is usually mounted under the dash, within easy reach of the driver, as shown in Fig. 5-6. This is not always practical, however, because of some factor in the automobile design. For example, the unit should not be mounted directly in the heater air stream, because temperature extremes could affect operation, and

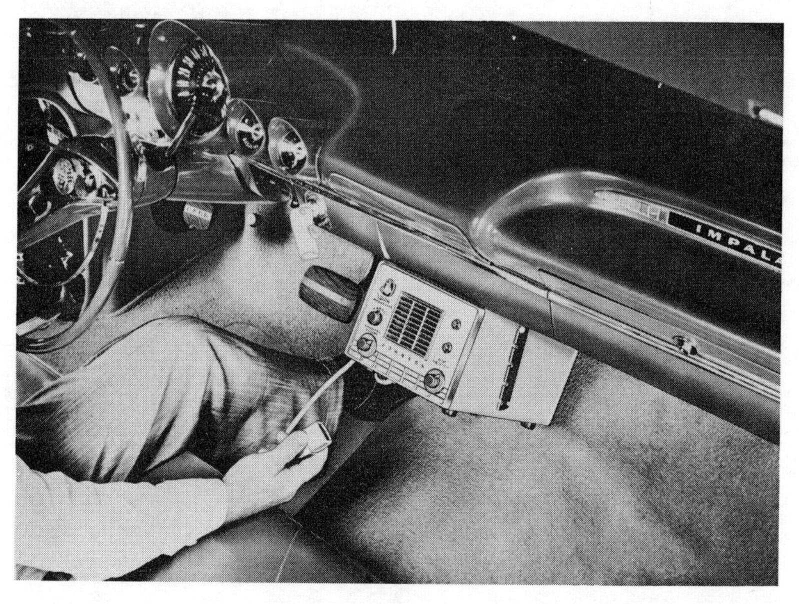

Courtesy E. F. Johnson Co.

Fig. 5-6. Typical mobile installation.

Fig. 5-7. One type of adjustable under-the-dash mounting brackets.

excessive heat can damage components. (Remember, too, that crystal mikes can be damaged by temperatures above 120°F.)

The transceiver must also be mounted in such a way that it does not interfere with proper operation of the vehicle. If it were located beneath the dash but too near the steering column, it could interfere with brake-pedal travel or cause the driver discomfort when applying the brakes. Equally dangerous is a mounting too far to the right, requiring the driver to lean over to reach the set. These are only a few of the possibilities that should be considered before choosing the mounting location.

To complicate matters, you will often find that the underside of the dash is either not level, or is cluttered with such things as heater and air-conditioner controls, auxiliary switches, cigarette lighters, ash trays, and the like. In such cases, there are several alternatives: (1) choose another location for the transceiver; (2) relocate the interfering object(s); or (3) use mounting brackets that will extend below the interfering object.

Fig. 5-7 shows a set of brackets which can be adjusted to different widths; loosening the wing nut on either side permits the unit to be positioned at various angles. Fig. 5-8 shows how longer brackets permit the radio equipment to be mounted directly below an obstacle mounted under the dash.

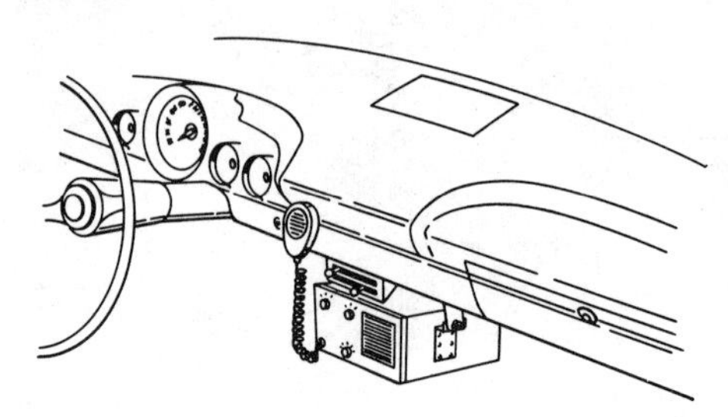

Fig. 5-8. Suitable brackets permit transceiver to be mounted below interfering objects.

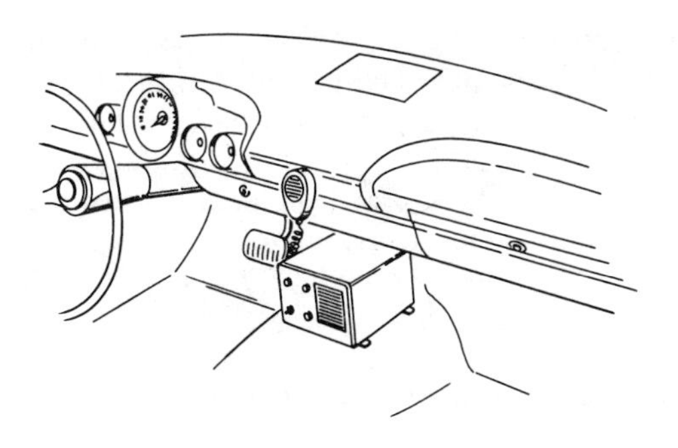

Fig. 5-9. A floor mounting can be used if no space is available under the dash.

If it is impractical to fit the transceiver under the dash, perhaps it can be mounted on the hump in the center of the front floorboard, as shown in Fig. 5-9. The microphone hanger can be fastened to the dash within easy reach of the operator. (Usually, the hanger on the transceiver can be removed and fastened to the dash with a couple of small sheet-metal screws.) Self-tapping screws are also generally used to fasten the transceiver to the floorboard, since the transmission located directly beneath makes the use of bolts and nuts impractical.

Antenna Cable

After mounting the transceiver and antenna, the next step is to figure out how the antenna cable is to be routed and what length will be needed. (It's always better to have a little too much than not enough.) The cable route should be as short as possible, at the same time away from the engine, gauges, switches, relays, etc.—which all tend to induce noise. If a rear-bumper antenna mount is used, the cable can be run through the inside of the car as shown in Fig. 5-10.

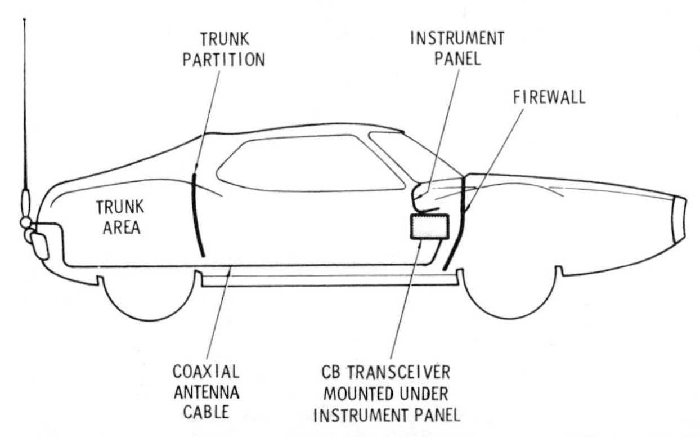

Fig. 5-10. One method of routing the antenna cable in a mobile installation.

On an inside route, lay the cable along the edge of the floor, beneath the floor mat or carpeting. The rear-seat cushion will have to be removed, and perhaps a small hole will have to be drilled, in order to feed the line from the antenna into the trunk area. Always check for existing openings before drilling a hole. If it does become necessary, fit the hole with a rubber grommet to prevent the cable from being damaged. If the antenna is mounted on the rear deck, the cable can be routed in the same manner. With roof-top antenna installations, however, you should follow the procedures provided with the antenna for routing the cable to the transceiver.

Power Source

CB equipment can be connected to the automobile power source at several points, as shown in Fig. 5-11. Practically all commercial transceivers include a protective fuse of some type, either in series with the "A" lead, or in a holder on the equipment itself. Locate the fuse before installing the equipment so you will know where it is, should trouble occur.

Keep the supply lead as short as possible. Wire has a certain amount of resistance which causes some of the source voltage to be

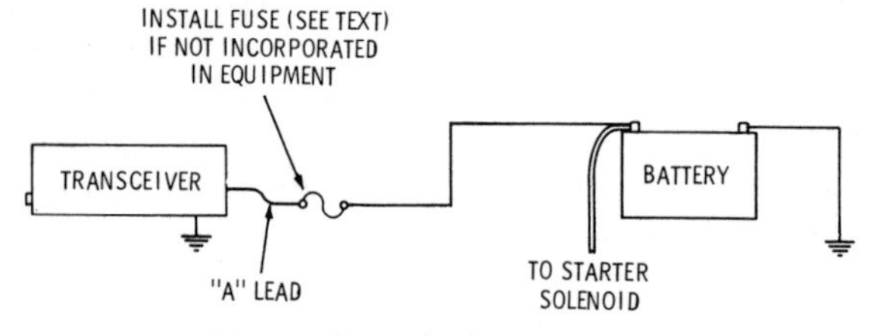

(A) Directly to the battery terminal.

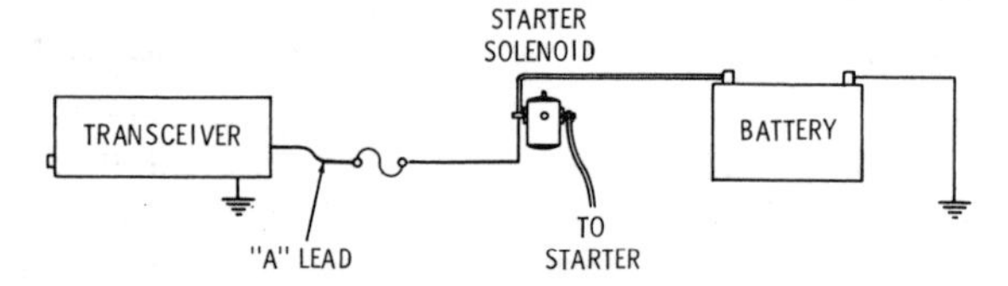

(B) Connection to the starter solenoid.

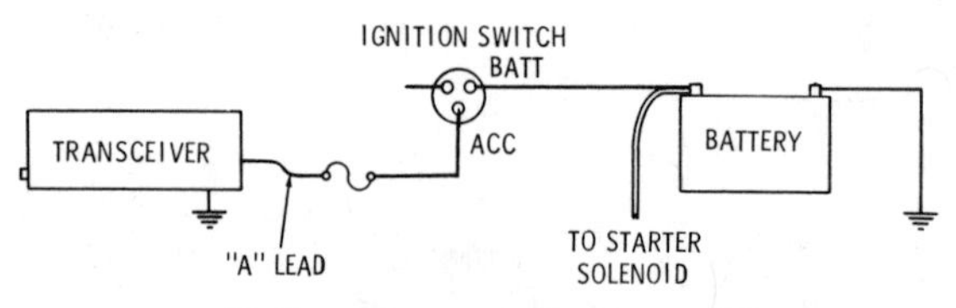

(C) Connection to the ignition switch.

Fig. 5-11. Three methods of connecting a transceiver to the automobile power source.

dropped across it. In Fig. 5-11A, the "A" lead is connected directly to the battery terminal. The dc path is completed from the battery through the metal framework of the car to the transceiver chassis.

The starter solenoid may be located closer to the radio equipment. If so, the "A" lead can be connected to its "hot" side, as shown in Fig. 5-11B. The solenoid can be located by tracing the battery cables; one will be connected to ground (the engine or frame), and the other will lead to the starter solenoid. Observe polarity when connecting the equipment to the source voltage.

Fig. 5-11C shows how the radio equipment can be connected to the ignition switch. The "A" lead can be connected to either of two terminals in this instance. If you wish the transceiver to operate only when the ignition switch is on, connect it to the accessory (ACC) terminal. The equipment can be used anytime (with the ignition switch on or off) if connected to the battery terminal. The former method can prevent a dead battery should the unit be unknowingly left on for any length of time.

Perhaps you have a transceiver that was originally purchased for 6-volt operation but is now to be installed in an automobile having a 12-volt electrical system. Previously we spoke of a voltage loss between the transceiver and the power source because of dc resistance in the wire. Now such a drop is desirable. The same connections shown in Fig. 5-11 can be used, but an additional resistance is placed in series with the "A" lead, as shown in Fig. 5-12. The value of this resistor is chosen so that approximately 6 volts will be dropped across it, thereby providing only the required 6 volts at the transceiver. The value is not given here since it will vary with power consumption; however, it can easily be computed by Ohm's law. Also, the amount of power to be dissipated will determine its wattage.

This arrangement cannot be used with all equipment, because the amount of voltage dropped across the resistor depends on the current through it and this current is not the same for the receiving and transmitting modes. Most transceivers draw more current when transmitting than when receiving. This, of course, means that a resistor chosen to drop 6 volts when receiving, will drop even more when transmitting. This will reduce the voltage at the transceiver to less than 6 volts, and transmitter power will be reduced accord-

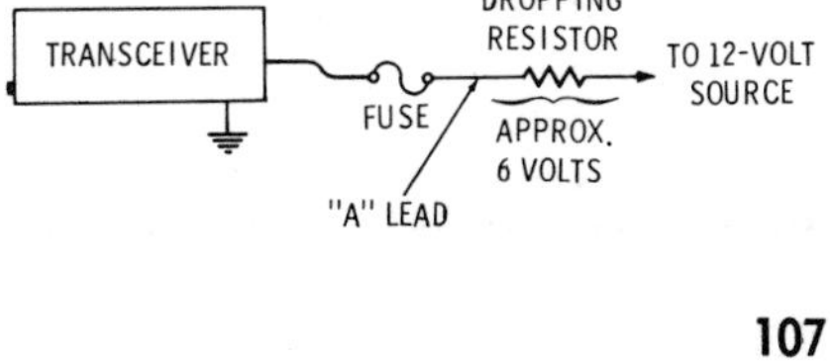

Fig. 5-12. Method of connecting a 6-volt transceiver to a 12-volt source.

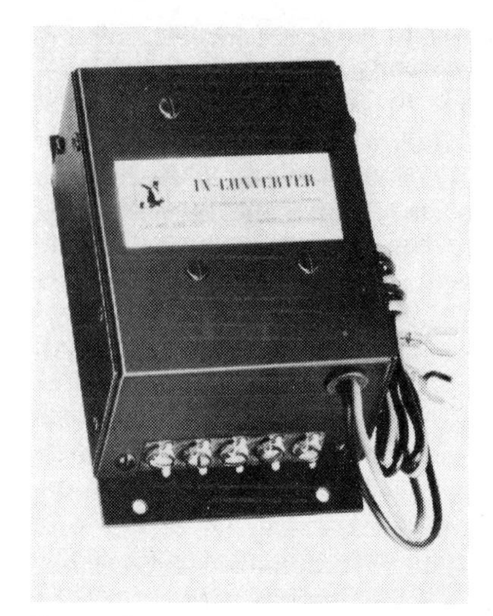

Courtesy E. F. Johnson Co.

Fig. 5-13. The "In-Converter" is designed to convert 6 volts dc to 12 volts dc and to provide polarity inversion.

ingly. Consult the service literature for the power consumption of the equipment in question. If the difference between transmitter and receiver requirements is significant, this method should not be used.

In almost all cases, power converters, like those shown in Figs. 5-13 and 5-14, are the best solution to the problem. If the installation

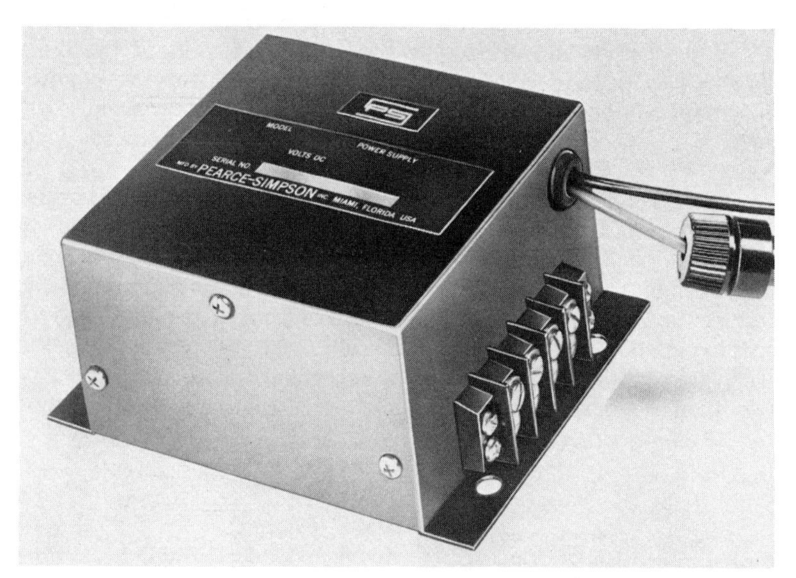

Courtesy Pearce-Simpson, Inc.

Fig. 5-14. Another example of a power and polarity converter for mobile installations.

requires connecting a 6-volt transceiver to a 12-volt source or a 12-volt transceiver to a 6-volt source, the units shown here will handle the job. Furthermore, both provide 12- to 12-volt polarity inversion, thereby permitting "negative ground only" equipment to be operated from 12-volt positive ground electrical systems and vice versa.

When installing a CB unit designed to operate from various power sources, be sure to use the proper terminals or the correct power cord, whichever the case may be. Also remember that the polarity of the voltage source is very important, especially with transistorized equipment. This equipment will not operate if connected in reverse, and may even be damaged.

Where polarity must be considered, it is important to know which terminal of the car's battery is connected to ground. Since the adoption of the 12-volt electrical system, almost all American cars use the negative-ground electrical system.

A transceiver designed to be powered only from 110 volts ac can also be used in a mobile installation with the use of an inverter, which converts dc into household current (110 volts, 60 hertz). Fig. 5-15 shows one type of inverter, a portable unit that can be plugged into the receptacle of a standard cigarette lighter. The ac voltage is available from the outlet at the far right. The control just to the left is a four-point voltage regulator that permits the voltage to be adjusted for minimum and maximum loads, as well as for variations in the dc input voltage. This particular inverter operates from a

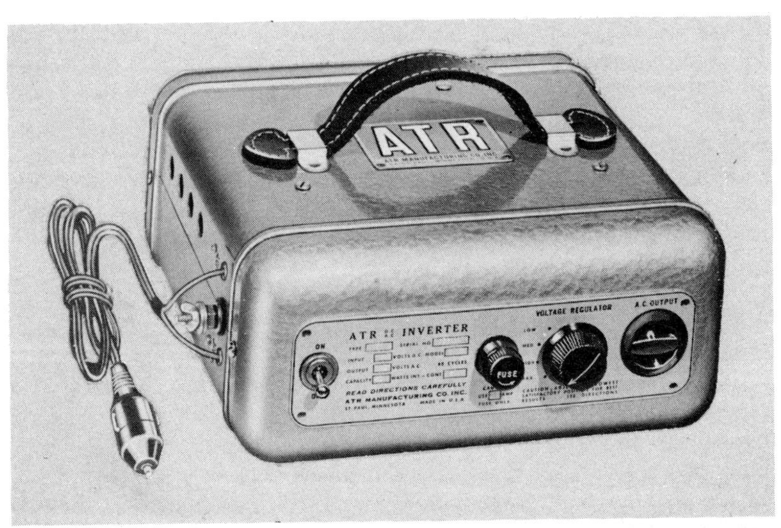

Courtesy American Television and Radio Co.

Fig. 5-15. The Model 12T-RME portable inverter.

12-volt dc source and provides a 110-volt, 60-hertz output at 90 watts continuous and 125 watts intermittent. This is more than adequate to operate the majority of Class-D transceivers, which usually require no more than 60 to 70 watts. However, if more power is required, larger models are available.

There are several possible locations for such a device when used in automobile installations. Here again, much depends on the type of car, size of unit, available space, etc. Smaller inverters can be set on the front seat or on the floorboard—or, if adequate space is available, perhaps mounted under the dash beside the transceiver. If not, there is usually sufficient room in the trunk. Suitable mounting brackets to securely fasten the inverter under the dash or in the trunk are available as accessories for permanent installations. Also available with some units is a small remote-control unit that fits under the dash for the trunk-mounted inverter, providing complete control from within the vehicle.

One of the most commonly asked questions concerning a two-way radio installation in an automobile is: "Will it run the battery down?" It depends on the power consumption of the equipment and how much it is used while the engine is idling or turned off. The batteries employed in modern cars are designed to supply a great deal of power. Although late-model broadcast receivers generally draw less current than their older counterparts (which did not have low-current tubes and transistors), additional current is required for such things as power windows, seat, windshield wipers, etc.

The average Class-D Citizens-band transceiver should cause no undue strain on the present electrical system of the car, provided the latter is in good working condition and the radio equipment is used with discretion. Like any other electrical device, it would not be advisable to operate the equipment for prolonged periods when the engine is not running.

If you have trouble keeping the battery charged, even under normal operation, it would be advisable to have the electrical system of the car checked, paying particular attention to the voltage-regulator setting. If most of your driving is confined to short runs in the city, for example, the voltage regulator can usually be adjusted to provide a higher charging rate. However, to prevent possible damage, there is a maximum setting which should not be exceeded.

If the present generator or alternator is not capable of safely providing enough current to keep the battery sufficiently charged, an oversized replacement may be the answer. Newer model cars are equipped with an alternator. This is an ac generator which operates in conjunction with a rectifier to supply dc current to the battery. One of the nice features of the alternator is that it is capable of providing a substantial charging current even at idling speed.

NOISE AND INTERFERENCE SUPPRESSION

Since the majority of pulse-type noises disrupt the amplitude rather than the frequency of radio signals, they have their greatest effect on a-m communications. Frequency modulation (fm) is relatively unaffected by most types of noise because of the action of the limiters. Noise-limiter circuits (discussed in Chapter 3) are employed in most Citizens band equipment; however, additional steps may have to be taken to suppress noise, especially in mobile installations. Don't forget, the noise level is one of the limiting factors in communicating range.

Noise Sources

Noise falls into one of two categories—natural or man-made. Lightning caused by atmospheric disturbances is one example of natural noise. Some of the sources of man-made noise that primarily affect fixed-station installations include electric mixers, hair dryers, fans, drills, fluorescent lights, and medical equipment, to name just a few. In vehicular installations (cars, boats, airplanes, etc.) the electrical and ignition systems are the chief sources of noise. Basically there are three ways in which noise can enter a transceiver: (1) through the antenna; (2) through the power source; (3) from radiation pickup by the internal circuit. Some types of noise can be overcome or minimized; others cannot.

Generator Noise

As mentioned previously, few American cars are being built with dc generators; most employ alternators. However, where generators are used, a serious noise problem often exists. Generator noise is especially troublesome at high frequencies. It can be recognized by a whine that varies in pitch with the speed of the engine. This noise is caused by arcing between the generator brushes and commutator. A bypass capacitor will usually minimize or completely eliminate this trouble. A standard capacitor (0.5 μF) may already be connected to the armature terminal of the generator to prevent interference on the regular broadcast receiver. However, this capacitor, because of its construction, has a certain amount of inductance which reduces its efficiency at higher frequencies. Therefore, a noninductive coaxial feedthrough-type capacitor should be connected in series with the wire going to the armature terminal of the generator as shown in Fig. 5-16. The case must be properly grounded to the generator; it may be necessary to scrape or sand the surface under the mounting screw to provide a good ground contact. The capacitor must not be placed in the field lead, since the generator could be damaged if the capacitor should short. If the noise is still severe after

Fig. 5-16. A noninductive capacitor installed on a generator.

this capacitor is installed, the trouble may be due to worn brushes or a dirty or worn commutator. If this is the case, the generator will have to be overhauled.

A device designed especially to suppress generator noise in the 14- to 30-MHz range is shown in Fig. 5-17. Electrically it comprises a high-Q parallel-tuned circuit. With the unit connected in series with the armature lead to the generator, the capacitor is tuned for maximum noise reduction at the frequency used.

Ignition Interference

The ignition system is another common source of noise. It produces a popping sound which is especially noticeable when the engine is running slow. It also increases with engine speed, but is easily distinguished from generator whine. Ignition noise is caused by high-voltage discharges at desired and undesired points in the system—for example, between high tension wires, from the wiring to some point on the engine, down the outside of a greasy spark plug, or across the gap of a poor connection. Ignition noise may not be due to defects in the system. A certain amount is produced during normal operation. This noise can often be reduced or eliminated by using resistor-type spark plugs or radio-resistance high-tension wire to interconnect the coil, distributor, and spark plugs.

Courtesy Globe Electronics

Fig. 5-17. A tuned generator-noise suppressor.

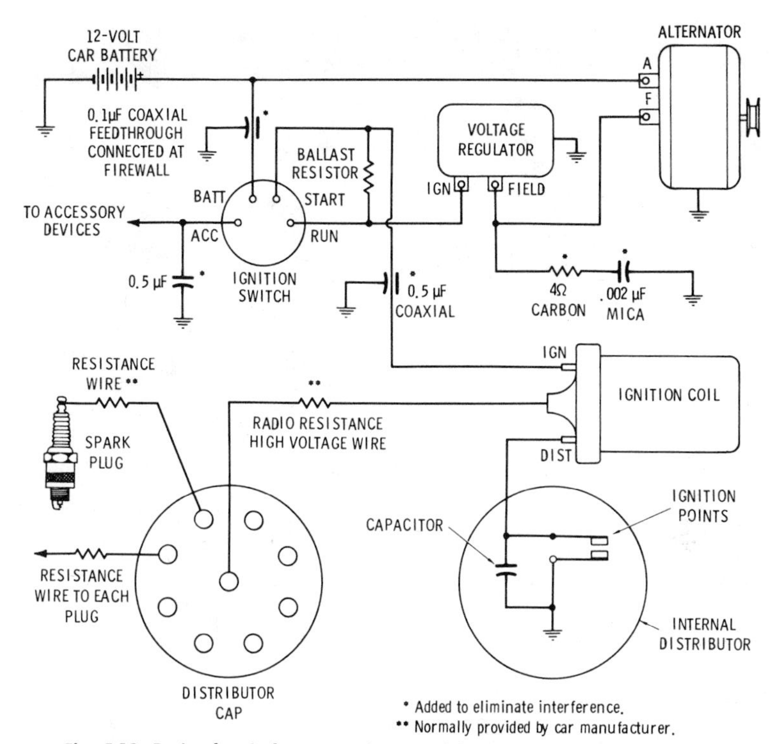

Fig. 5-18. Basic electrical system of automobile showing noise-suppression devices installed.

Do not, however, use both radio-resistance wire and resistor spark plugs, as the efficiency of the ignition system will be impaired. Fig. 5-18 shows the basic electrical system of an automobile and indicates components that can be installed to minimize or eliminate interference. Complete noise-suppression kits similar to that shown in Fig. 5-19 are available from several commercial sources. These kits usually include all appropriate suppression devices and wiring, plus installation instructions.

Other Interference

Another type of noise is produced by the voltage regulator. This "hash" type interference occurs as the relay points open and close. It can usually be eliminated by installing an RC network at the field terminal as shown in Fig. 5-18.

Other sources of noise are the gas gauge (both the sending unit at the tank as well as the gauge on the panel), the turn-signal flasher, and the brake-light switch. Noise produced by these sources can be eliminated in most cases by connecting a noninductive by-

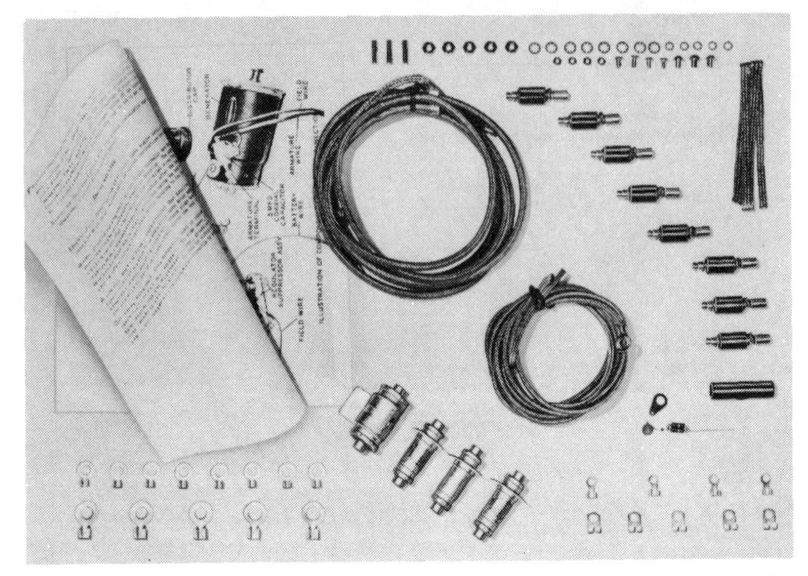

Courtesy E. F. Johnson Co.

Fig. 5-19. Typical noise-suppression kit.

pass capacitor in series with the "hot" lead to each unit.

Static electricity creates still another type of noise. It, too, is recognized by a popping sound from the speaker, but unlike ignition interference this noise will generally continue as long as the vehicle is in motion and even with the engine turned off. In fact, this is one way of identifying it.

Whenever metal joints in an automobile body become loose, there is a chance that a static electric charge will build up between them. When this charge reaches a certain value it will discharge as a spark to adjacent metal structures, thereby causing a popping sound to be heard from the receiver. The best-way to prevent or eliminate static interference is by bonding together any metal

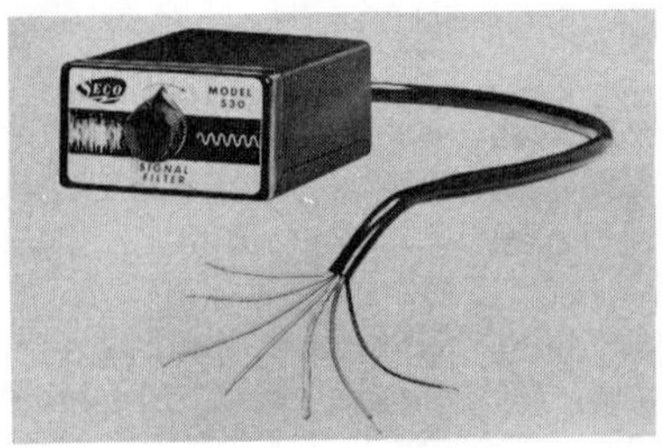

Fig. 5-20. The SECO Model 530 "Signal Filter", used to eliminate ignition interference, hash, and background noise.

Courtesy SECO Electronics, Inc.

structures that are now loose or could become loose. A bonding strap may have to be used to bond the instrument panel to the frame or the hood and trunk lid to the body or frame; and in some cases the fenders may even have to be bonded to the frame before the trouble can be eliminated.

Stations operating at fixed locations such as a home or office are also troubled by noise, but not to so great an extent as in automotive installations. Noise is often picked up through the ac power line and transferred into the transceiver circuits. A line filter connected in series with the power cord will generally correct this situation.

A number of commercial devices designed to suppress radiated man-made noise and interference are currently available for use with CB equipment. Fig. 5-20 shows one example. These devices can be used for either mobile or fixed station installations.

6

Maintenance and Repairs

Citizens-band units, like any other man-made equipment, occasionally require the services of a skilled technician. Many repairs will be routine, involving nothing more than replacing a tube, transistor, diode, or similar component; others will be more complex, calling for the troubleshooting techniques of a specialist.

According to the FCC regulations, not just anyone can service two-way radio equipment. Any transmitter tests or adjustments which may affect the legality of operation must be made by, or under the immediate supervision of, a person holding either a first- or second-class commercial radiotelephone license. Changing adjustments, disturbing the chassis wiring, or replacing critical components may cause off-frequency operation, introduce spurious radiation, or increase transmitter power beyond the legal limit. Obviously, before an unlicensed person can perform any type of tests, repairs, or adjustments, he must first know what changes can affect proper operation. This is one place where the old saying, "a little knowledge can be a dangerous thing," holds true.

Notice that the regulations state "by or under the immediate supervision of a properly licensed person." This means that such tests and adjustments can be made by anyone, as long as he is supervised by a person with the proper license. The licensed person is then held responsible for the proper functioning of the equipment after such tests, repairs, or adjustments have been completed.

Certain exceptions to the rule make it possible for an unlicensed person to construct, install, and service certain commercially manufactured Class-C and -D equipment. For example, no commercial radiotelephone license is required to construct a kit or to install and maintain other commercial equipment in which the frequency-determinating elements of the transmitter have been preassembled, pretuned, and sealed at the factory. Replacement or adjustment of any components which might cause off-frequency operation cannot be made without first breaking the seal (see paragraph D, Section 95.97, of the FCC regulations in the Appendix). If a technician intends to service CB equipment, he must obtain a commercial license.

TEST EQUIPMENT

Many shops already have most of the test equipment needed to service CB equipment, and will thus need only a few additional items. A good vom with a sensitivity of at least 20,000 ohms per volt and a vtvm are already included on most shop benches.

Although most tubes will be checked by substitution, a tube tester can be used to confirm leakage, shorts, low emission, and other types of tube defects, some of which are not always readily apparent. Transistors can be checked for leakage, open elements, or shorts with an ohmmeter or a transistor tester.

When circuit alignment becomes necessary, as it will in any radio equipment of this type, a stable signal generator (preferably crystal calibrated) is required. It should be able to cover all rf and i-f ranges from 150 kHz to 30 MHz, be capable of delivering a signal of less than 1 microvolt, and have provision for signal modulation. An additional generator may have to be acquired if Class-A equipment is to be serviced.

Some type of dc power supply will be needed at the bench when equipment is brought into the shop for repair. The dc source can be a fully charged storage battery capable of delivering sufficient current, or a battery eliminator. A power supply that delivers fixed dc voltages (both 6 and 12 volts), or a filtered variable dc supply such as that used to service transistorized equipment can be employed. The latter is more desirable since it will deliver either 6 or 12 volts, and can be varied both above and below these amounts to simulate actual dc voltage variations encountered during normal operation. A dc power supply of this type is shown in Fig. 6-1.

In addition to the basic test equipment just mentioned, a couple of other instruments not usually included in the average service shop will also be needed. One of these is an rf wattmeter for checking the transmitter output signal. The second, and one of the most important as far as legal operation is concerned, is the frequency meter. It, of course, is needed to check the operating frequency of the transmitter, which should be a routine part of every service job. Several instruments are suitable for this purpose, one of the most popular being the heterodyne frequency meter.

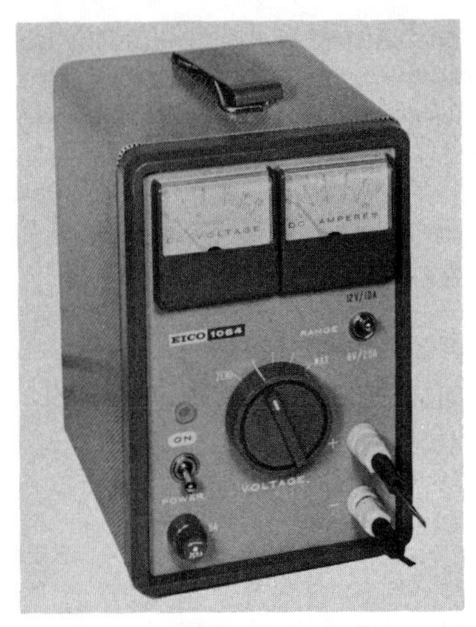

Courtesy EICO Electronic Instrument Company, Inc.

Fig. 6-1. A variable 6/12 volt filtered dc power supply.

Test instruments designed primarily for checking the operation of Class-D Citizens band equipment are available. Among other things, these instruments check modulation, rf power output, antenna efficiency, field strength, etc. There are also several all-purpose CB testers like the ones in Figs. 6-2 and 6-3 which are practical for the CB'er in making routine checks. Among other functions, the test set in Fig. 6-2 can be used as an rf wattmeter, swr (standing wave ratio) bridge, modulation monitor, field-strength meter, and crystal checker. The unit in Fig. 6-3 serves as a signal monitor, crystal-con-

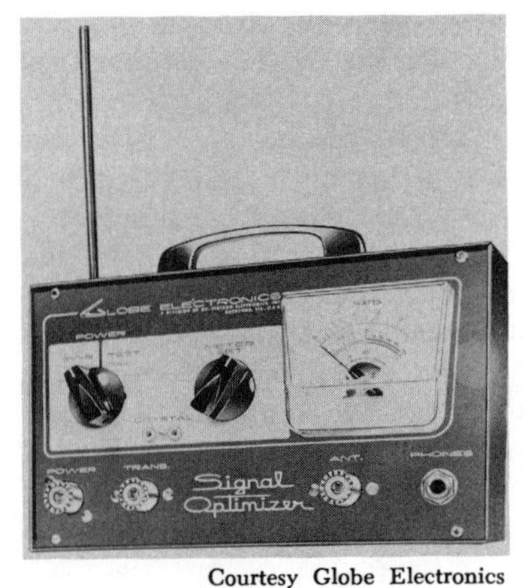

Courtesy Globe Electronics

Fig. 6-2. The Globe "Signal Optimizer" CB tester.

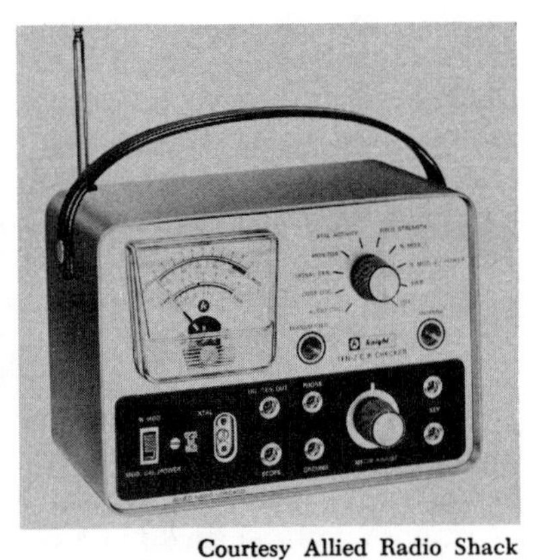

Courtesy Allied Radio Shack

Fig. 6-3. The Knight-Kit "Ten-2" portable CB checker.

trolled signal generator, and audio oscillator. In addition, measurement of swr, rf power output, modulation percentage, field strength, and crystal activity can be made with it. Other test equipment for checking the operation of CB radio equipment is described in Chapter 7.

SERVICING HINTS

When a piece of equipment becomes defective, check the service manual first for possible causes. Not only can this save time, but it will often prevent additional troubles caused by promiscuous probing and testing in an attempt to locate the trouble.

Lead Dress and Component Placement

Improper lead dress, especially in the rf and i-f stages, can cause unstable operation and/or result in spurious oscillations. When a component is to be replaced, the new part should be located and wired exactly like the original. Replacement components should always be of the same value and rating; also, it is not wise to try to improve on the original design.

Temperature-Compensated Components

Components used in critical circuits, such as the oscillator, may be temperature-compensated. Replacement with a standard type can cause improper operation. The value of a negative temperature-coefficient capacitor varies inversely with temperature. In other words, as the ambient temperature increases, the capacitance decreases, and vice versa; thus, the circuit in which it is used will be relatively unaffected by heat. Fig. 6-4 shows the location of temperature-compensated capacitors in a typical transmitter oscillator circuit (Fig. 6-4A) and a receiver oscillator circuit (Fig. 6-4B). Circuits are not all alike; however, this will give you a general idea of where this type of component is used. It should not be difficult to determine if a defective capacitor is temperature-compensated or not. One with a negative-temperature coefficient, for example, is designated by the letter *N* followed by several numbers, such as N150. If it is not stamped on the capacitor, it will most likely be indicated in the parts list.

Printed-Circuit Boards

Printed-circuit boards have replaced the once conventional "hand-wired" circuits in most modern CB equipment. Fig. 6-5 shows an example of such a board. Usually, printed-circuit boards are made of a phenolic material. The wiring pattern, which is a flat metal foil conductor, is imprinted or etched on the board, and is sometimes

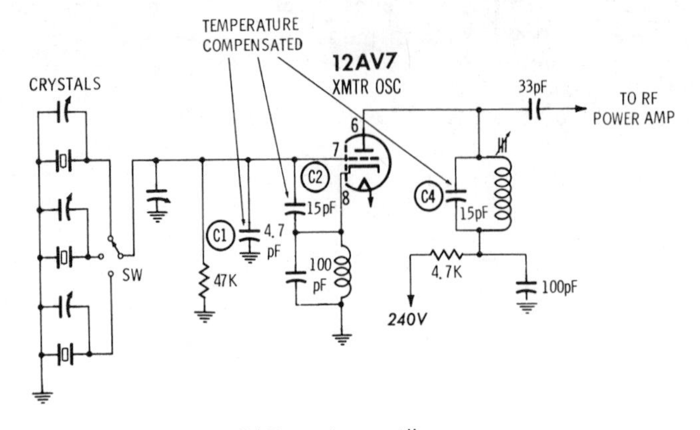

(A) Transmitter oscillator.

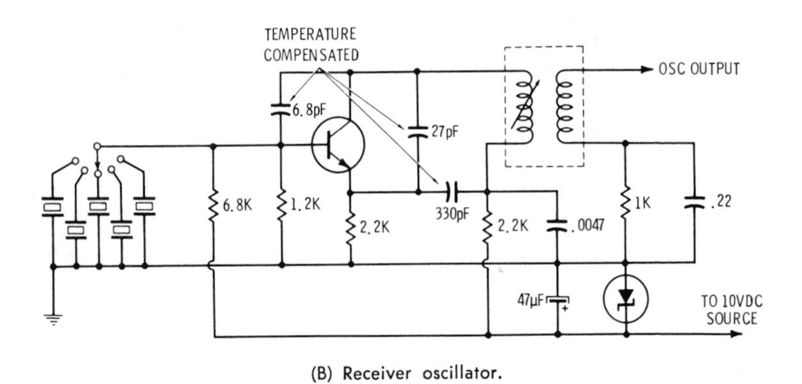

(B) Receiver oscillator.

Fig. 6-4. Location of temperature-compensated capacitors in typical transmitter and receiver rf oscillator circuits.

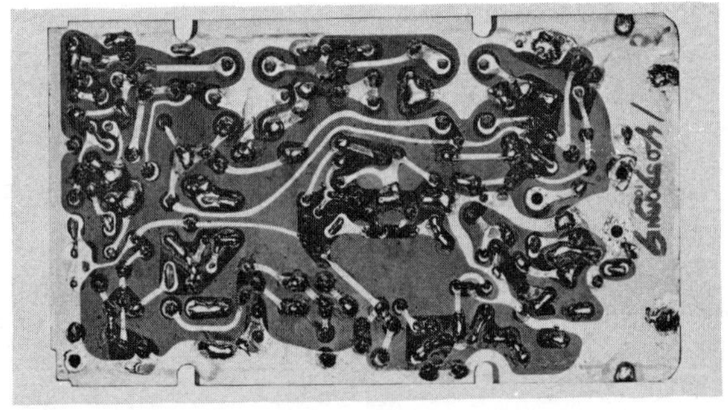

Fig. 6-5. A typical printed-circuit board.

covered with an epoxy resin to prevent dust and moisture from affecting operation. The circuit components, which may be located on either or both sides of the board, are connected to the printed wiring through holes in the board. This method of construction eliminates the old-style array of wires and provides a more compact assembly. However, special care must be exercised when servicing equipment of this type, since these boards can be easily damaged. Occasionally one of the conductors will break due to physical strain, rough handling during servicing, etc. Often this will be nothing more than a hairline break, although it can be sufficient to disable the entire transceiver. If a break should occur, try to locate it without removing the circuit board from the chassis. Sometimes it can be spotted by placing a light on the underside (opposite the wiring side) of the board. Although this board is flexible, avoid bending it in order to prevent damaging the printed wiring.

When checking a printed-circuit board, it's best to use a test prod with a sharp point to pierce the epoxy resin coating over the printed wiring. Tests should be made at soldered junctions rather than by punching holes at just any point along the delicate conductors. (Such holes represent potential breaking points.) In a conventionally wired circuit, it's common practice to unsolder one lead of a component before testing it. With printed-circuit boards, however, special soldering techniques must be used to prevent damaging the board. Therefore, check carefully and be reasonably certain a part is defective before unsoldering it. Then use a low-wattage soldering iron, since the bond between the printed wiring and the board can be broken by excessive heat. Also, using too much solder on a connection can result in a short between conductors.

Transistors

Transistors and other solid-state devices, like printed circuits, also require special considerations when being tested and replaced. Basically there are three types of transistors: (1) the small low-power type for rf and i-f circuits; (2) the medium-power audio transistor; and (3) the high-power transistor used in audio-output stages and power supplies. Some of the most frequent defects in these units are leakage, shorts, and open elements. In some circuits, like the one in Fig. 6-6, the emitter resistor (R1) is fusible and will occasionally open, "killing" circuit operation. When replacing a fusible resistor, make sure the replacement has the same characteristics as the original resistor had.

The schematic symbol in Fig. 6-6 is commonly used to represent the transistor. A transistor in an integrated circuit is represented by the same symbol but without being enclosed in the circle. Fig. 6-7 shows several methods of identifying the transistor terminals. The

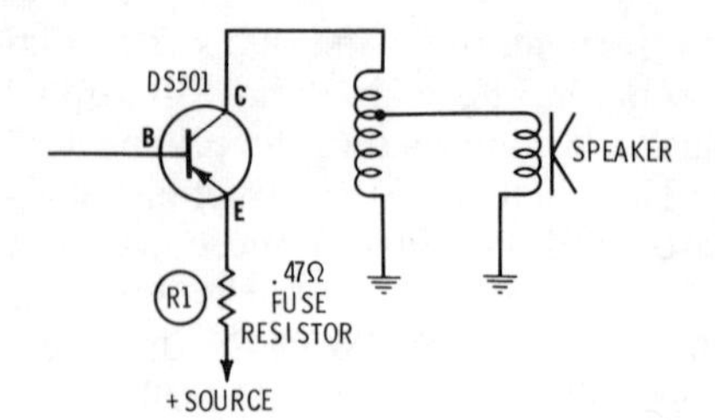

Fig. 6-6. Schematic symbol used to represent a transistor.

leads in Fig. 6-7A are unevenly spaced, and the collector lead is isolated from the other two. In Fig. 6-7B the leads are evenly spaced, but the collector is marked with a bright red or orange dot. The two terminals of the power transistor in Fig. 6-7C are marked "E" and "B." The only remaining connection is the collector—in this instance, the transistor case itself.

(A) Uneven spacing of leads. (B) Even spacing; colored dot identifies collector. (C) Pins marked with letters.

Fig. 6-7. Methods of identifying transistor leads.

It is not absolutely necessary to acquire a transistor tester just for CB servicing. If one is already on hand—fine! For the most part, though, a vom will suffice. If a high-impedance dc voltmeter is available, it can be used to make in-circuit tests. If it becomes necessary to remove a transistor from the circuit, several tests can be made with an ohmmeter. Fig. 6-8 shows how the front-to-back resistance ratio can be checked. For a low-power transistor, the ohmmeter is set on the R × 100 scale and the test leads are applied to the base and emitter (Fig. 6-8A). The reading is noted, then the leads are reversed. The same measurements are made between the base and

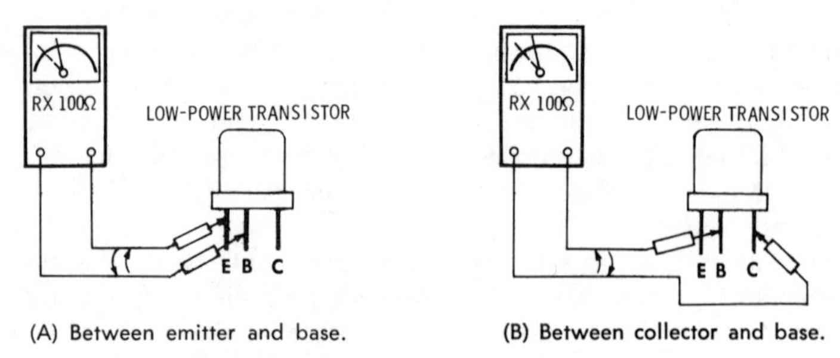

(A) Between emitter and base. (B) Between collector and base.

Fig. 6-8. Checking the front-to-back resistance ratio of a transistor.

collector terminals, as shown in Fig. 6-8B. In either case, a very high reading will be obtained in the reverse direction. In fact, the needle may not even move from the peg. In the forward direction, however, the reading will usually be less than 1000 ohms for the low-power type. Failure to obtain a reading in either direction indicates an open element. An extremely low reading in both directions signifies leakage or a short between the transistor elements.

Power transistors can be checked in the same manner, except that the ohmmeter must be set at the R × 10 range. The high-resistance reading of power transistors should be 5000 ohms or higher, while the forward reading will be approximately 100 ohms and may be even less.

Considerable heat is generated within a power transistor and must be dissipated to prevent it from being damaged. Some type of heat sink is used for this purpose. It may be an insulated portion of the transceiver chassis, or a separate device. When replacing a power transistor, make sure it is mounted firmly on its heat sink to provide maximum transfer of heat. Make it a point to always check these mountings when a piece of equipment using power transistors is serviced.

Small, low-power transistors generate very little heat. In fact, they are usually cool to the touch. Excessive heat at the leads when soldering one into a circuit, however, can easily cause damage. Therefore, always grasp the lead being soldered with a pair of long-nose pliers, between the transistor and the soldering iron. The pliers will act as a heat sink and prevent possible damage. Plug-in transistors should be removed from their sockets before heat is applied to the socket terminals. Always use a low-wattage soldering iron in transistor circuits.

Caution should also be exercised when troubleshooting solid-state circuitry with an ohmmeter as the voltage applied to the test leads of some meters can exceed the ratings of the components being checked. Thus, circuit damage or component failure could result. Also, you should avoid indiscriminately grounding various terminals throughout the circuit, and be certain of your test points.

LOCALIZING DEFECTS

Before a defective transceiver is removed from a permanent installation, a check of all logical trouble possibilities may save lots of time. A disconnected power cord, a blown fuse, or possibly even a defective power cord or power receptacle could be the reason. Some units are equipped with a pilot light that indicates when power is applied. This can be helpful in determining if supply voltage is present.

In vehicular installations, make certain the battery voltage is up to par and the fuse (located either in the "A" lead or in the transceiver) is not defective. Don't just look at it. Visual inspection of fuses is often deceptive, since a hairline break in the element may not be detected with the naked eye. Fuses should be checked with an ohmmeter, or replaced. If the replacement fuse does not blow, chances are the transceiver is not at fault. A hairline break is not always a sign of trouble in the radio equipment. Often it is caused by physical stress or by supply-voltage surges. Repeated fuse failures of this type, not traceable to the CB equipment itself, may be due to trouble in the electrical system of the car, possibly a misadjusted voltage regulator. This is further evidenced by frequent burnout of not only the tube filaments but also various lights throughout the vehicle.

If the fuse is found to be good but the equipment still does not receive power, check the "A" lead and the connection to the voltage source. Also check the equipment for proper ground.

A transceiver could be receiving power but still not transmit or receive signals. The first points to check in such instances are the antenna connectors, the antenna itself, and the transmission line. You will find cases where the antenna cable has been pinched under the edge of the rear seat, or where there is low source voltage due to a cable half-eaten away by battery acid. You will also come across antenna and "A" leads that have literally been sawed in two by windshield-wiper mechanisms, loose connections, and poor grounds of all kinds.

Only after all possibilities of this type have been checked, should the transceiver be removed from its mounting. Even then, it's best to try to repair it on the spot, where its antenna and source voltage are available. In this way, repairs and adjustments can be checked under actual working conditions. Some repairs, of course, are impractical to handle in the field, in which case the unit should obviously be taken into the shop.

Adjustments of some type will be a part of practically every service job, and a check of the operating frequency should be a matter of routine. This chapter, however, is concerned only with physical defects; servicing adjustments are discussed in the next chapter.

Isolating the Trouble to a Section

After removing the equipment to be serviced, you will have to determine which section is defective—the transmitter, receiver, or power supply. It's unlikely that all three will become defective at one time, although a common receiver-transmitter stage may develop trouble and impair the operation of these two sections.

Furthermore, since the power supply is common to all stages, a defect in it may affect the operation of all three sections of the transceiver.

Persistent fuse blowing is caused by excessive current being drawn by the equipment. This could mean trouble in either the power supply, transmitter, or receiver circuits. If a universal supply is employed and the unit operates normally on either ac or dc, but not on both, a power-supply defect is indicated. However, should the fuse blow on both types of operation, the trouble probably lies within the transmitter or receiver circuits.

Before checking the unit, disconnect the antenna (where applicable) and attach a dummy antenna in its place to prevent interfering with other stations while testing. Commercial dummy antennas (also called dummy loads) are available for this purpose, or a dummy antenna can be made by connecting a No. 47 pilot lamp across an extra antenna connector, as shown in Fig. 6-9. (This same method can also be used with other types of antenna connectors.)

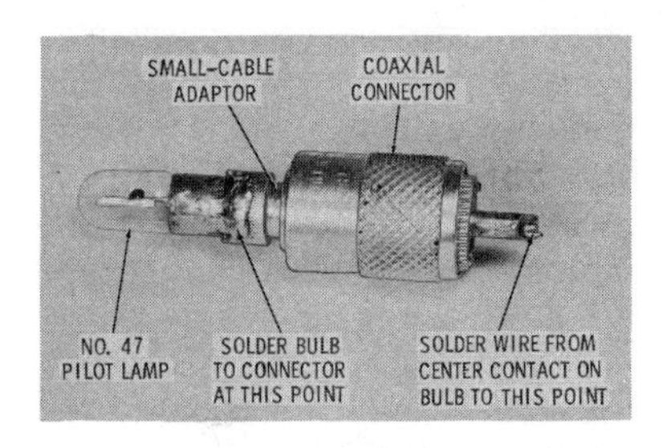

Fig. 6-9. A dummy antenna load constructed from a PL259 connector and a No. 47 pilot lamp.

Check to see if the tubes light. If so, depress the transmit button and notice whether the pilot lamp used as the dummy load lights. It should if the transmitter is radiating rf energy. If not, determine whether the receiver is working by turning the volume control fully clockwise and rotating the squelch control to maximum in both directions. A rushing sound should be heard from the speaker. Negative results on both tests indicate possible trouble in the power supply. This can be confirmed by a voltage measurement in the power supply. Suspect an open circuit, possibly the power switch, or an open choke coil or B+ resistor. A short in the power supply would normally cause repeated fuse blowout.

In a case where the receiver alone is working, a simple test with an additional receiver will often prove helpful in determining the condition of the transmitter. To perform this test, temporarily disconnect the dummy load and connect a regular antenna in its place (one should always be available at the bench). Tune or switch the extra receiver to the frequency of the transmitter in question, and turn the squelch to the Off position. The usual background noise will now be heard. Now momentarily depress the transmit button on the defective unit and speak into the microphone. If the background noise is silenced on the test receiver but no voice is heard from the speaker, the rf carrier is being produced and radiated, but there is no audio modulation of the signal.

The transmitters used in CB transceivers are relatively simple and should not be too difficult to troubleshoot. By using a little common sense, you can often save much time in a case like this. For example, here we have a receiver that works properly, which means the audio-amplifier and output stages (which usually act as the speech amplifier and modulator when transmitting) are working all right. The fact that rf energy is being radiated is reason enough to assume that the transmitter oscillator (and rf power-amplifier stage if used) are functioning—at least to some degree. Because of the continuous flexing of the microphone cord, one of the inner conductors will occasionally break. This, of course, eliminates the audio from the carrier, and could easily be the case here. To check, hold the transmit button down and speak into the microphone while twisting and bending the cable. This should cause the voice to be heard intermittently at the receiver. A different broken connector in this same cable could have prevented the relay change-over system from operating, allowing the receiver to continue operating when the mike button is depressed. Other causes for loss of audio modulation are defective components between the mike input and modulator stage (coupling capacitors, resistors, microphone preamp stage, etc.) or the microphone itself could be defective. If practical, it is a good idea to try another microphone.

Assuming the trouble occurs during the dc operation and a vibrator power supply is used, connect the unit to a variable dc power supply having appropriate meters to read voltage and current. Switch it to the proper voltage range and turn the transceiver on. Notice the amount of current drain. If it's excessive, reduce the input power to a safe level. (Check the service literature to determine what the normal current drain should be.) Turn the power off, disconnect one lead to the buffer capacitor(s) located across the secondary of the power transformer, then momentarily apply power again. If current drain is reduced appreciably, adjust the power-supply control until the rated voltage is applied and note the current reading. If normal, turn off the equipment and replace the defective buffer capacitor. (Make sure the replacement is of the correct value and voltage rating.) If two are used, it would be wise to replace them both. Do not operate the equipment for any length of time with the buffer(s) disconnected. Other than tubes, a defective buffer capacitor is one of the most common vibrator power-supply troubles, second only to the vibrator itself.

If the buffer(s) are checked and found to be good, remove the vibrator and replace it with one known to be good. Turn on the power supply and again check the current drain. If it returns to normal, the original vibrator is defective. (The reason for checking the buffer capacitor first is that a new vibrator may otherwise be damaged.)

Occasionally you will find that both the vibrator and buffer are bad. It's a good habit to always replace the buffer(s) when a new vibrator is installed. In fact, many manufacturers won't guarantee their vibrators unless this is done.

Transistor power supplies present considerably fewer problems than vibrator supplies. The fact that the vibrator operates on the mechanical switching principle makes it understandable that failure is bound to occur. One solution to vibrator replacement is to substitute an electronic switching device similar to the one shown in Fig. 6-10. Known as the Perma-Vibe, this unit contains a transistorized power supply oscillator that is constructed so that it can serve as a direct replacement for most 12-volt vibrators.

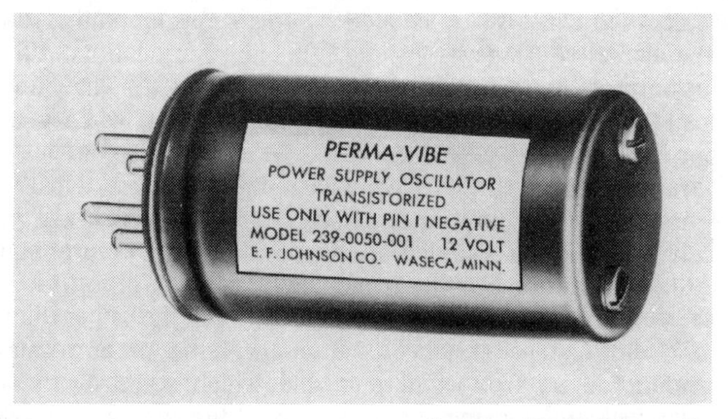

Courtesy of E. F. Johnson Co.

Fig. 6-10. A transistorized power-supply oscillator designed to serve as a direct replacement for most 12-volt vibrators.

In tube-type equipment, the majority of equipment malfunctions will be due to tube defects. Open filaments can often be spotted by a quick visual inspection. Other tubes defects can be readily located by direct substitution (the preferred method) and can usually be confirmed by a tube tester. Unlike most electronic equipment, many CB units will not work properly if one or more tubes become weak. Therefore, it will often be necessary to replace a tube that would ordinarily give much longer service in other types of equipment.

Finding the Defective Stage and Component

After determining which section of the equipment is defective, standard techniques can be applied to further localize the trouble. In the receiver section, for example, a signal generator can be used to inject a modulated signal at the grid and plate of each stage, working backward from the audio output to the antenna as shown in Fig. 6-11. First disconnect the antenna. Then, with the transceiver turned on, inject a signal at the grid of the output tube. If this stage is operating normally, a tone will be heard from the speaker. Next, apply the generator signal to the plate of the audio amplifier. The tone should become increasingly louder as each additional stage adds amplification. When you get to the input of the detector, switch the generator to the i-f frequency, and to the rf frequency when you reach the grid of the converter. When you get ahead of the inoperative stage, the signal will no longer be heard. If you received a normal response at the input of the audio detector, but heard nothing at the grid of the second i-f stage, the trouble lies somewhere between the grid of the second i-f and the audio detector. With the aid of a schematic and a vom, a few voltage and resistance measurements should reveal the defective component with little trouble.

A weak stage can also be located by this same method. However, when the defective stage is passed, the only indication will be a less-than-normal increase in gain. The converter stage normally produces little if any stage gain, so bear this in mind when you evaluate your findings.

The transmitter in the average Class-D transceiver usually consists of no more than three or four stages, two of which are part of the receiver circuit. This leaves only one or two more stages—namely the transmitter oscillator and the rf power output. Occasionally, the rf power-output stage is not used; on the other hand, an additional audio stage serving as a microphone preamp may be. There should be no trouble determining which stage of the transmitter is defective by using the process of elimination. Here again, the defective component in the suspected stage can be isolated by the use of standard voltage and resistance measurements.

Make a visual examination of the components in the defective stage. Often a charred or cracked resistor, or a blackened wire or burned spot on the chassis, will immediately indicate the faulty part. However, don't just replace a burned component; check further to determine why it burned. Your sense of smell can be of help, since most parts give off a rather unpleasant odor when they become overheated. Moreover, placing the tip of your finger on the body of a large wirewound resistor, which ordinarily should get very warm, will sometimes reveal an open resistor. Listening will often pay off

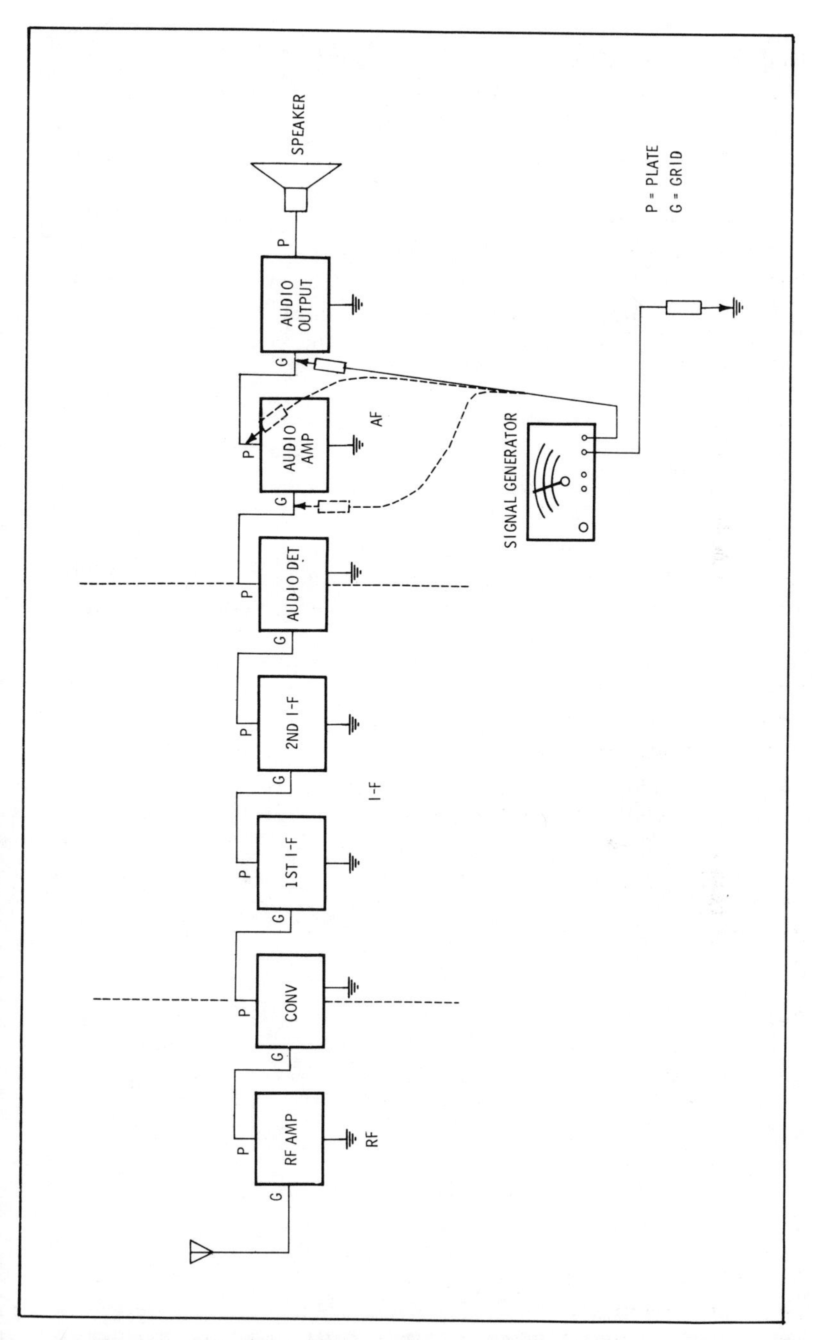

Fig. 6-11. Method of localizing a defective stage by signal injection.

as well. When the transceiver is turned on, listen for the buzz of the vibrator (in dc sets). If a jar is required to shock the vibrator into operation, the contacts are probably sticking and it should be replaced. Likewise, one that has a rough or erratic tone which changes pitch intermittently should also be replaced along with its associated buffer capacitor(s).

The basic procedures for isolating defects in solid-state equipment are essentially the same as in tube-type models. Only the method of checking components is different and certain precautions (some of which have already been explained) must be observed. Transistors themselves normally require less attention than tubes.

PREVENTIVE MAINTENANCE

Preventive maintenance can keep equipment from breaking down at inopportune times. By pulling the chassis out of the cabinet at least every few months and performing a few simple operations you may save yourself or your customer future trouble.

To check the unit over completely, it must be fully accessible. If the unit is a tube type, check all tubes and replace any that are weak, leaky, or shorted. Check power-supply voltages and make certain all crystals are plugged in tightly. If the push-to-talk relay contacts are accessible, clean them with carbon tet or other grease-dissolving solvent, and wipe them dry with a soft cloth. Rub a clean strip of paper between the contacts to burnish them. (Never use sandpaper, emery cloth, or other abrasive.) Brush all dirt and dust from the chassis. Finally, check the transmitter output and, if necessary, adjust for maximum efficiency. Any transmitter tests made with the dummy antenna disconnected should be kept as brief as possible.

7

Servicing Adjustments and Measurements

One of the most important parts of any service job is a check of equipment performance and adjustment of any circuit components for maximum efficiency. This is particularly true in low-powered CB equipment where all of the available transmitter power and receiver sensitivity is needed.

A 5-watt plate input power to the final transmitter stage will generally produce an output of somewhere between 2 to 3.5 watts. A Class-D transceiver capable of delivering 3.5 watts of output power is operating at 70 percent efficiency, which is considered very good for this type of equipment. This output is determined primarily by the efficiency and adjustment of the final stage and is affected to a large degree by associated circuit adjustments.

Periodic adjustments and frequency checks are therefore vital to proper operation, and should be made at least every six months during normal usage, every three months in commercial applications, and any other times deemed necessary or when repairs are made. No service job is complete without checking the transmitter frequency and the input and output power of the final stage, as well as making the necessary adjustments to provide a maximum transfer of power between the transmitter and the antenna, and a check of receiver operation. Some of these tests may have to be performed before or during a service job, but they should definitely be repeated as a final step after all repairs are completed. A simple thing like replacing a defective tube or transistor in the transmitter oscillator or rf power-output stage is sufficient reason for making frequency adjustments.

CHECKING TRANSMITTER OPERATION

Only persons who hold a first- or second-class radiotelephone license are permitted to make tests and adjustments that affect transmitter power and frequency. Before making any adjustments on the transmitter, remove the antenna lead and attach a dummy load in its place. A No. 47 pilot lamp, connected as shown in Chapter 6, is suitable for indicating that the transmitter is radiating energy and that modulation is taking place. However, a better impedance match will be obtained by connecting two 100-ohm, 2-watt resistors in parallel in place of the lamp. More effective dummy loads than either of these can also be purchased. Any adjustments or tests made without such a device should be kept as brief as possible. A simple field-strength meter (Fig. 7-1) is helpful in determining whether a transmitter is "putting out." It also provides a relative indication of the signal field strength.

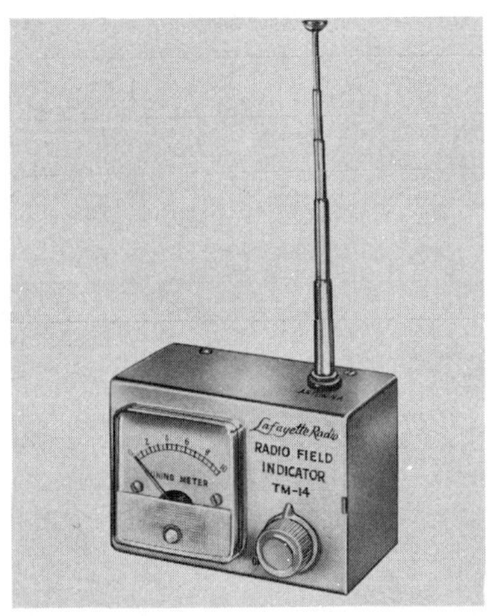

Courtesy Lafayette Radio Electronics.

Fig. 7-1. Model TM-14 field-strength indicator.

Operating Frequency

The most important, and surely the most frequent, measurement involves checking the transmitter frequency. Several instruments are available for this purpose; the accuracy of the type chosen will depend on the class of CB gear to be checked. The most critical, of course, are Class-A units which must operate within a frequency tolerance of 0.00025 percent. This calls for an extremely accurate frequency meter—one with an even closer tolerance. Class-D stations, with a .005 percent tolerance, are considerably less critical.

A frequency meter used to check Class-D equipment should be accurate within at least .0025 percent.

One instrument commonly used for this purpose is the heterodyne frequency meter. With this device, a signal from a built-in tunable oscillator is heterodyned with the transmitter signal to obtain the measurement. Another commonly used meter indicates the difference between the frequency of the transmitter and a crystal-controlled reference frequency within the instrument.

A tolerance of .005 percent means that a Class-D station operating on a frequency of 27.005 MHz cannot deviate more than ±1350 Hz per second. In other words, if the frequency checks between 27.00365 and 27.00635, it is within the legal limit.

Before checking the transmitter frequency, turn on both the transceiver and the frequency meter and allow sufficient time for the equipment to reach operating temperature. (This will ensure an accurate reading.) If a heterodyne frequency meter is to be used, make sure it is properly calibrated and set up for the test as described in the instruction book. Then depress the mike button and adjust the dial of the meter to produce a zero beat with the transmitter signal. The transmitter frequency is then determined from the setting of the meter dial or from the calibration chart. Meter accuracy must also be taken into account when evaluating the reading. The meter itself should be accurate to a much higher degree than the tolerance of the frequency to be measured. Always let the transmitter operate a minute or two while measuring the frequency. Some crystal-oscillator circuits have been known to perform satisfactorily at first, only to jump off frequency after a few seconds or so of operation.

Just because a crystal is guaranteed to be accurate within a specified tolerance does not mean it cannot operate off-frequency. Some crystals are designed to be used in a particular circuit, and will operate off-frequency if placed in another. Likewise, an improperly adjusted circuit can cause the same trouble. The crystal may also have been tampered with; if so, it should be replaced. Crystals are assembled, sealed, and tested at the factory, and no attempt should be made to open them. If the operation of any crystal is questionable, replace it. Fig. 7-2 shows a test set that is designed especially for checking crystals of the fundamental and overtone types. In addition, it can serve as an rf voltmeter or wattmeter when used in conjunction with an accessory rf probe.

Most crystals are designed to operate at their specified frequency when working into a capacitive load. The amount of this load has been standardized at 32 pF, which may be a physical capacitor acting in combination with stray capacitance and the input capacitance of the oscillator, or it may consist of the latter combination

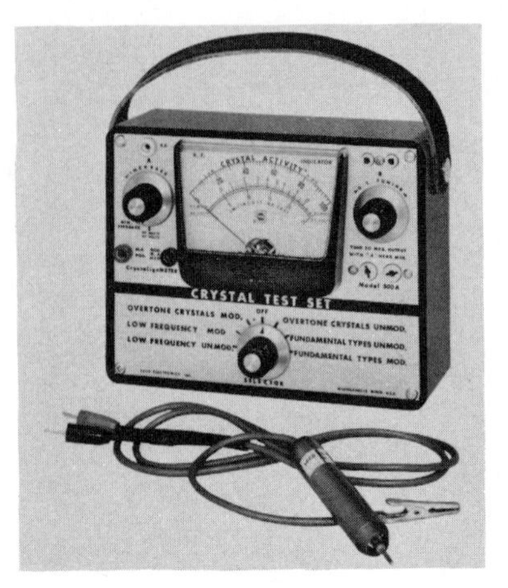

Courtesy SECO Electronics, Inc.

Fig. 7-2. The SECO Model 500A crystal test set.

alone. Fig. 7-3 shows a basic crystal-oscillator circuit that operates on the third mechanical overtone principle. The parallel-resonant circuit at the plate is tuned to the desired frequency. An overtone crystal designed to operate at antiresonance into an unloaded grid circuit is used here. Capacitor C1, shown by the dotted lines, represents the stray circuit capacitance and the grid-to-cathode capacitance. Replacing this crystal with one of a different type could easily cause off-frequency operation. In circuits where a fixed capacitor is used across the crystal, you will generally find it's less than 32 pF. The value of this capacitor, however, in conjunction with stray and input capacitances, should add up to the needed 32-pF load.

The oscillator circuit will usually have an adjustable trimmer capacitor across the crystal to provide a means of varying the capacitive load and subsequent frequency of the oscillator stage. In multichannel oscillators (containing more than one crystal), a trimmer may be placed across each crystal or a single trimmer may be used for all crystals. The operating frequency must be checked on each channel used. The frequency adjustment for the crystal(s) installed in the equipment is made at the factory; however, if addi-

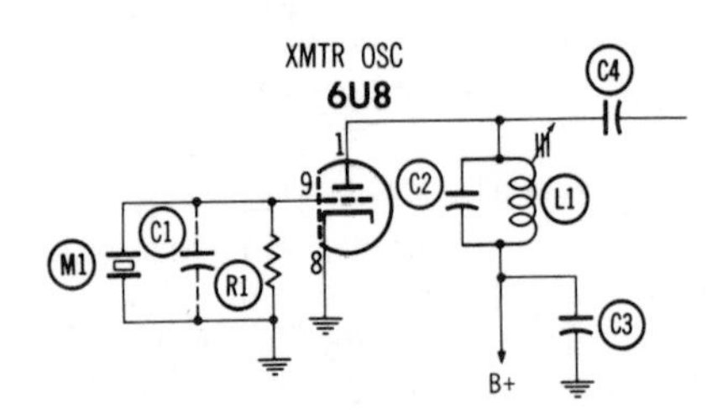

Fig. 7-3. Crystal-oscillator circuit.

tional crystals (even of the proper type) are added or one is replaced, the operating frequency of each channel should be rechecked.

In a fixed-tuned single-channel unit, all adjustments are made at only one frequency. In multichannel equipment, however, these adjustments are usually a compromise between the various channel frequencies. This compromise is best reached after checking and adjusting the operating frequency of the transmitter, by making the rest of the adjustments on a center channel frequency. In other words, if Class-D Channels 9, 11, and 13 are used, the transmitter adjustments should be made with the selector in the Channel-11 position.

To further guarantee legal operation, check the frequency while varying the source voltage both above and below the rated value. The frequency of a unit designed to operate at 115 volts should remain within legal limits at voltages ranging from 90 to 130 volts. Battery-operated sets should be checked from 5 to 7.5 volts for a 6-volt unit, and from 10 to 14 volts for 12-volt sets.

Power Input

Another important factor to consider when checking the transmitter is the amount of input power to the final rf stage. This stage may be the rf power amplifier, or the transmitter oscillator itself in equipment where the former circuit is not used. Naturally, full use should be made of the permitted 5 watts input; however, this amount cannot be exceeded. In tube-type sets, power input is computed by multiplying the plate voltage by the plate current. To obtain these figures, measure the dc plate voltage with a vom. Then connect a milliammeter in *series* with the plate lead, and note the amount of current. Let's assume the plate voltage is 250 volts and the current is 20 milliamperes (or .02 ampere). Input power would then be exactly 5 watts—the legal limit. The FCC regulations state that the average power input to the plate (in tube-type equipment) or collector circuit (in the case of transistorized units) or circuits which contribute radio-frequency energy to the antenna, cannot exceed the limits specified by the regulations for the applicable class of service.

Transceivers rated at 5 watts input may not always provide this amount in actual practice. If the power is too low, a close check of the equipment will usually reveal a below normal source voltage, power supply defect, weak final rf tube or defective transistor. In this type of equipment, the circuitry must operate at peak efficiency in order to obtain the maximum output (within legal limits).

Modulation and RF Output

After installing a CB transceiver, one of the first procedures involves checking to see that maximum power is being delivered to the antenna. A dummy load, consisting of a No. 47 pilot lamp connected across the antenna terminals, can be used to indicate the presence of rf energy. Speaking into the microphone should cause the brightness of the lamp to vary in accordance with voice peaks if the transmitter carrier is being modulated.

Although power output can be indicated by a pilot light, an rf wattmeter with a 50-ohm impedance is more desirable. Two types of rf wattmeters are commonly used for this purpose. One not only measures the amount of rf power, but acts as the dummy load as well. Hence, it is referred to as a termination-type meter. The other type is connected between the transmitter and the dummy load (or the antenna) to indicate the amount of power being delivered. Instruments of this type for commercial service shops are generally too expensive for the average CB'er to purchase. However, some testing devices are available within the average budget.

Fig. 7-4 shows a CB transmitter tester that can be used to measure either relative rf power output or modulation percentage. As an rf

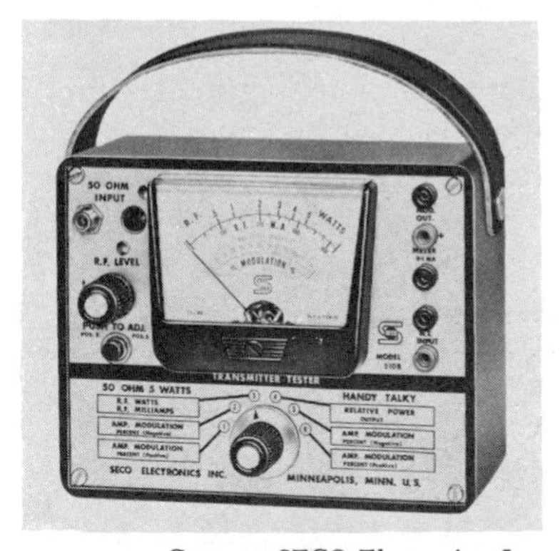

Courtesy SECO Electronics, Inc.

Fig. 7-4. SECO Model 510B CB transmitter tester.

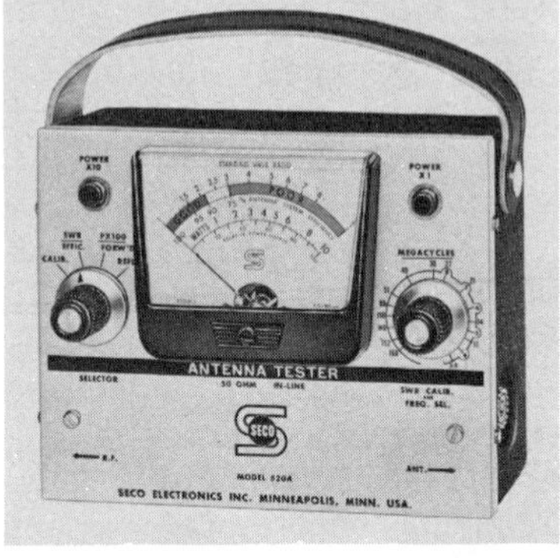

Courtesy SECO Electronics, Inc.

Fig. 7-5. SECO Model 520A calibrated antenna tester.

wattmeter it functions as one of the terminating-type instruments; the rf energy is dissipated in a six-watt 50-ohm resistive load. Modulation percentage is read directly from the lower scale on the meter, and either positive or negative peaks can be measured.

The antenna tester shown in Fig. 7-5 is another useful instrument. It has a meter with scales calibrated to read relative forward and reflected power in watts, swr, and antenna system efficiency. This tester is connected in series with the coaxial line between the transmitter output and the antenna.

Fig. 7-6A shows an in-line, or series-type, rf meter used to check impedance matching between the transmitter output and the antenna

(A) Antenna meter calibrated to read mismatch in SWR.

(B) "CB Matchbox" designed to reduce or eliminate impedance mismatch.

Courtesy E. F. Johnson Co.

Fig. 7-6. Two devices used to insure proper impedance match between transmitter and antenna system.

system. It provides relative indications of forward and reflected rf power, and has a scale calibrated directly in swr (standing wave ratio). This type of meter must be terminated with either an antenna or dummy load of the proper impedance. Swr readings provide an indication of the degree of impedance mismatch between the transmitter output and the antenna. Instruments such as these may be left in the line indefinitely because they introduce only a negligible amount of insertion loss. The associated device in Fig. 7-6B contains a matching network that can be adjusted to provide the impedance match needed to ensure the maximum transfer of rf energy from the transmitter to the antenna system.

When making transmitter adjustments, connect the test equipment as recommended by the equipment manufacturer and perform the circuit adjustments according to the procedures given in the service manual. These procedures will differ for various pieces of equipment.

CHECKING RECEIVER OPERATION

One of the most important factors of receiver operation is sensitivity. No matter how efficiently the transmitter operates, the communication link is broken if the other party cannot be received.

Reduced sensitivity is a common trouble; however, it usually occurs at a gradual rate and often goes unnoticed. Three common causes include improper supply voltages, a change in component values due to a defect or gradual aging of components, and weak tubes or defective solid-state elements. Another probability is circuit misalignment due to exposure to excessive moisture or indiscriminate tampering with tuning adjustments.

Receiver alignment will generally be no more difficult than for most standard a-m broadcast receivers and some of the simpler units will be even easier. Dual-conversion superhets will perhaps be a little more difficult because there are two i-f frequencies to contend with.

8

Operating Procedure

The number of Citizens-band stations is increasing by the thousands each month. Such expansion can only mean that many of the CB channels will become quite crowded—especially those in the Class-D service. Thus, it is important that every operator become thoroughly familiar with the proper equipment usage—not only in accordance with FCC regulations, but also with courtesy and respect for others who must share the same channels. If each person would follow the correct procedures outlined within this chapter, much of the congestion caused by improper, lengthy, and unnecessary transmissions would be reduced to a minimum. This, of course, would benefit everyone.

HOW TO OPERATE CB EQUIPMENT

After receiving your station license, obtaining the desired CB equipment, and making the necessary installations, you are just about ready to begin operations. First of all, however, it is important that you have a basic understanding of how to operate the equipment.

One attractive feature of the Citizens-band transceiver is its simplicity of operation. Although there is a variety of Citizens-band equipment on the market, the number of basic operating controls varies only slightly from unit to unit. Some equipment, because of more elaborate circuits or special features, will have more controls than others.

Volume Control

The controls employed on the Class-D transceiver in Fig. 8-1 are typical of those found on the majority of transceivers in this class.

The volume control is one of the basic controls used on all CB equipment. As the name implies, its purpose is to control the volume level —but only of the receiver output, not the audio level of the transmitter as some would think. On some equipment, the volume control may be a step-type arrangement—a rotary switch with three or more positions. The desired volume level is selected by switching to the position that most nearly meets the demand. This switch, however, cannot be set to obtain a level between these positions. The most popular control used in present-day CB equipment is the continuously variable resistance type consisting simply of a potentiometer which is used to adjust the volume to any desired level within its range of rotation.

Courtesy Dynascan Corp.

Fig. 8-1. The Dynascan "Cobra V" is a solid-state transceiver with a typical complement of operating controls.

Power Switch

The on-off power switch on most CB equipment is mechanically affixed to the rear of the volume control and works in conjunction with it—just as on most radio or TV sets. Turning the control clockwise switches the unit on and increases the volume. Conversely, turning it counterclockwise reduces the volume and, at maximum rotation, actuates the switch that turns the unit off. Many CB units employ a pilot light to indicate when the equipment is on. Occasionally, you may find a unit that has the power switch separate from the volume control. This will usually be a two-position toggle switch or one of the slide type.

Squelch Control

The squelch circuit is incorporated in practically all late-model Citizens-band equipment. This control adjusts a cutoff voltage that the incoming signal must overcome before it can be heard from the speaker. When the squelch control is backed off, the speaker will reproduce the usual random background noise similar to that heard from the average broadcast receiver when tuned between two stations. When a signal of significant strength is received it will "override" much or all of this background noise. When the transmission ends, the background noise will again emanate from the speaker. By advancing the squelch control to the point where the receiver is silenced, nothing will be heard from the speaker until a signal of sufficient strength is again received.

Any signal that lacks the strength to overcome the squelch will not be heard. This is just as well, since in most cases it would be unintelligible anyway. The further the squelch control is advanced, the stronger the signal must be to reach the speaker.

As a general rule, the squelch control should be adjusted to the point where the receiver is silenced and then just a bit further. This is referred to as the "threshold" setting. In mobile CB operation, it will probably have to be adjusted slightly beyond this point; otherwise, noise pulses from power lines, neon lights, factory equipment, etc., may be of sufficient strength to "break" the squelch. As you become familiar with the equipment, you will be able to determine the proper setting with little difficulty.

Channel Selector

There are three basic types of CB equipment—fixed-tuned, tunable, and a combination of both. In all cases, the transmitter is fixed-tuned for immediate operation on one or more channels; however, the receiver may be fixed-tuned with switch selection of one or more predetermined frequencies, or continuously tunable over a band of frequencies. In some equipment the receiver is fixed-tuned for operation on several channels and also tunable across the entire band.

Fixed-Tuned—The frequency-selection control on the multichannel fixed-tuned transceiver is usually labeled "Channel." This control will probably be a rotary-type switch with letters or numbers designating the various positions. Some fixed-frequency equipment may have provision for only two channels, in which case a toggle, slide-type, or two-position rotary switch may be used. The single fixed-frequency unit, of course, will not have any frequency-switching control. The operating frequencies of the fixed-tuned multichannel units can be selected by turning the selector to the desired channel position. Furthermore, the frequency at any of the switch positions can be altered by changing the crystals within the equipment. Therefore, even a single-channel unit can be made to operate on other channels with little difficulty. For this reason, the selector positions are usually marked with arbitrarily selected letters or numbers rather than with the actual channels or frequencies (see Fig. 8-1). The channel frequencies for the various positions depend on which crystals were provided by the manufacturer, and which crystals were added or changed after the unit was purchased.

Tunable—In this type of unit, the transmitter frequency is fixed but the receiver can be continuously tuned, usually over the entire band. This equipment, unlike the fixed-tuned, will have some type of dial indicating the relative setting for various frequencies. The desired channel to be received is tuned in much the same manner as a regular broadcast receiver.

Channel Selection

Although 23 Class-D channels have been allocated for CB use, the FCC has invoked a number of restrictions on their use. For example, Channels 1 through 8 and 10 through 23 may be used for communications between units of the same station, and Channels 10 through 15 and 23 are the only channels on which units of different stations can communicate.

Channel 9 is reserved solely for emergency communications where the immediate safety of lives or the protection of property is involved, or where communications are necessary to render assistance to a motorist. The licensee must determine ahead of time that his communications on this channel will meet one or more of the foregoing requirements.

Several other factors should also be taken into consideration when selecting a CB channel for operation. As you know, Citizens-band frequencies are shared with other units in this service. Therefore, try to pick a channel that is not too crowded, and at the same time one that will afford the least interference. For example, the FCC does not guarantee Class-D operators any protection from interference due to industrial, scientific, and medical equipment operating in the 26.96- to 27.8-MHz band. Much of this operation is on the frequency of 27.120 MHz. This frequency is located between Channels 13 and 14, so if you intend to use your equipment near hospitals or factories, it would be best to select a channel other than 13 or 14, and possibly not even 12 and 15 if the interference is too severe. Since Channel 23 is shared with Class C (for remote control of devices by radio) and other services, it may become quite congested and impractical to use in some localities.

The frequency chosen for operation will depend largely on the area where the equipment is to be used. A particular channel may be overcrowded in one area and practically unused in another. If in doubt as to which channels are best suited for your particular locality, ask some of the local CB operators.

In some areas, certain channels have been set aside (by gentleman's agreement) as *calling* and *working* frequencies. The calling frequency is used solely for contacting other stations. After communication is established with the desired station, both stations then switch to one of the "working" channels to exchange messages.

Other Control Features

In addition to the basic controls previously mentioned, some transceivers have a switch that permits the automatic noise limiter, squelch, or avc systems to be disabled. Other transceivers incorporate a signal spotting feature which enables the operator to accurately preset his receiver dial on the frequency on which he expects a reply. For example, if the operator of a transceiver having a tunable receiver wishes to call a station on 27.115 MHz (Class-D Channel 13) and expects a reply on this same frequency, he merely has to set the transmit selector to the Channel 13 position and actuate the spotting switch (this may either be a selector-type switch or a push button). He may then "pinpoint" this frequency on the receiver by adjusting the tuning dial until a peak is indicated by the S-meter; other frequency spotting arrangements might require tuning for a zero beat. Without this feature, inaccurate tuning could result in calls being missed. (Remember that by necessity a CB transceiver must tune quite "sharp" in order to separate stations.) To "spot" channels other than the one on which you wish to call, you merely set the transmit selector to the appropriate channel, "spot" it on the receiver dial, and then return the transmit selector to the desired operating channel before calling the other station.

Some transceivers also include a selective calling feature which accounts for at least one additional control, and there are several other controls or switches that will be found in some of the more sophisticated transceivers. It would not be practical, however, to go into a detailed discussion of these controls at this time since the primary concern here is with basic operating controls.

Operating the Microphone

There is a wide variety of microphones (discussed in Chapter 2) for use with Citizens-band equipment. Most of these microphones employ push-to-talk buttons, but some depend instead on a push button or switch (usually an automatic-return type) on the equipment itself.

The push-to-talk button is pressed to activate the transmitter. At the same time the receiver is automatically silenced. Releasing it disables the transmitter and at the same time restores the receiver to normal operation.

It was pointed out earlier that the power output of the transmitter is at a fixed level. This is the unmodulated rf carrier output in Fig. 8-2A. The voice spoken into the microphone produces an electrical equivalent of the sound waves (Fig. 8-2B). When fed into the transmitter, the voice signal is superimposed on the carrier signal as shown in Fig. 8-2C. This process is known as modulation.

When the press-to-talk button is actuated, a signal is being transmitted even though no words are spoken into the microphone. An unmodulated carrier (Fig. 8-2A) is being radiated from the antenna. Not until you speak into the microphone, however, is it modulated.

To produce a good, clear signal at the receiver, the microphone must be held the proper distance from the mouth. Holding it too close will cause the voice to sound garbled or muffled, and too far away will produce weak audio at the receiver. Best results will be

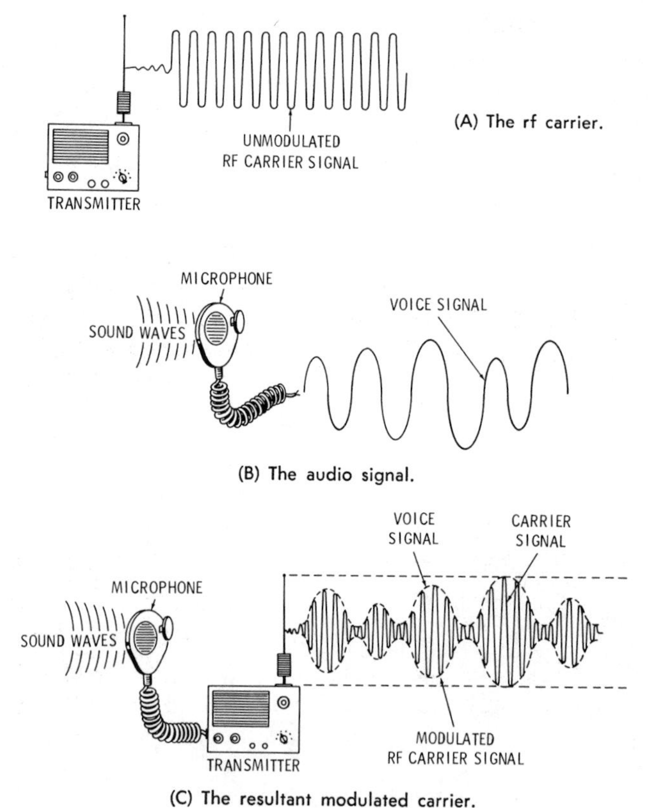

(A) The rf carrier.

(B) The audio signal.

(C) The resultant modulated carrier.

Fig. 8-2. Producing an amplitude-modulated carrier.

obtained by holding the microphone one to three inches from the mouth.

Occasionally, a whisp may be heard by the receiving station due to breath entering the microphone from close range. This situation can be relieved somewhat by holding the microphone farther away or by holding it at a 45-degree angle when speaking.

Even though the unmodulated rf carrier power of the transmitter is fixed, the level of the reproduced voice signal at the receiver will vary in accordance with the percentage of carrier modulation. The maximum modulation level allowed a CB station is 100 percent. The percentage of modulation increases when one is speaking loudly into the microphone, and decreases when speaking softly. This does not mean, however, that you should yell into the microphone. True, the higher the percentage of modulation, the louder the received voice signal; nevertheless, there is a limit. Fig. 8-3 better illustrates the effect of modulation on the transmitter carrier. Fig. 8-3A shows an unmodulated carrier. In Fig. 8-3B the rf carrier is modulated 50 percent. This is the equivalent of speaking into the microphone in a moderate voice. Fig. 8-3C shows a carrier that has been modulated 100 percent by a higher-level signal. Should the carrier be over-modulated (over 100 percent as shown in Fig. 8-3D), it would not only distort the voice signal, but also introduce harmonics which

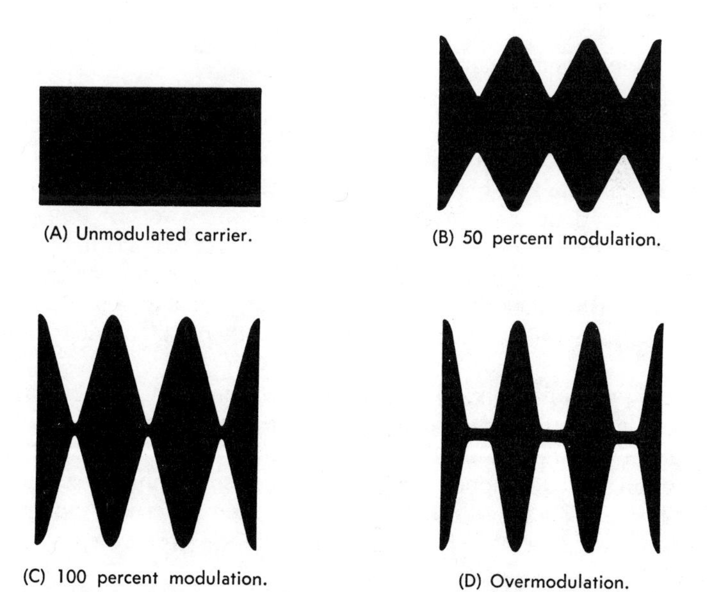

Fig. 8-3. The effect of modulation on the rf carrier.

could extend outside the assigned band and interfere with other services.

Since higher-modulation percentages extend the transmitting range to some extent, it is desirable to operate close to the maximum limit. Most Citizens radio equipment is designed to operate near 100 percent, but usually includes some type of self-limiting arrangement to prevent exceeding this amount.

STATION OPERATION

The FCC regulations discussed throughout this book were in effect at the time of publication; however, as the number of CB stations increases, revisions and extended control may become necessary. Indications at present point to further strengthening of these regulations.

The station licensee is solely responsible for the proper operation of his equipment. If he allows other persons to operate the station, he must assume full responsibility for their actions. Obtaining the license is relatively easy. However, you can lose your station authorization by operating contrary to FCC regulations. In fact, the FCC is currently imposing heavy fines on violators in addition to (or instead of) revoking the station license. To protect this privilege, you should become familiar with and operate according to the regulations governing proper operation. It is also necessary to review these regulations from time to time.

Permissible Communications

Citizens radio is intended to fulfill a definite need in connection with business or personal activities. It can be used for a personal convenience, such as establishing a means of communication between your automobile and home, or for practically any legitimate purpose. The applications are practically unlimited; however, transmitters should not be put on the air unless there is something that must be said. The CB channels are similar to a telephone party line and must be shared with other operators. Unnecessary transmissions serve only to cause congestion, thereby depriving others of needed channel space.

Many individuals are abusing the privilege granted them by operating their equipment in a manner contrary to good practice. Some are under the impression that Citizens radio can be used for experimentation or as a hobby, while others discuss sports, weather, or other trivia. This is not the intended purpose of Citizens radio. Used properly, it can be a useful communications tool for everyone; abuse it and you may be denied its use.

A CB station may communicate with operators in any of the other CB classes except Class C. It, of course, is for control purposes only and may not employ radiotelephone emission. Perhaps you are wondering how communications are carried on between different classes, since they are separated by other bands. Actually, you will be concerned with two types of operation—crosschannel and crossband. Either is easily accomplished if you have the proper equipment. If the frequency of your transmitter and receiver can be changed separately, you can engage in crosschannel communications with other operators having similar equipment. For example, if the controls at one Class-D station are adjusted to transmit on Channel 10 and receive on Channel 14, communications can be established with another Class-D station which can transmit on Channel 14 and receive on Channel 10.

Crossband transmission is practically the same, except for the receiver. Communication between Class-D and -A operators, for example, can be established if the Class-D operator has a Class-A receiver and the Class-A operator has a Class-D receiver. Inasmuch as each operator transmits only on the frequencies within the respective band for which he is licensed, no regulations are violated. The FCC is concerned only with transmitting frequencies.

Although channel and band jumping are permitted between CB stations, communications with operators of other services, including government and foreign stations, is prohibited. Some exceptions to this rule are made for operation during emergency situations (refer to Section 95.85 of the FCC rules and regulations in the Appendix).

In any case, transmissions should be made only when necessary, and even then they should be limited to the minimum practical transmission time. A radiotelephone code used for some time by other services has been adopted by many CB operators because it greatly reduces the time required to transmit a message. (See Table 8-1.) The following sequence will demonstrate how effectively it simplifies messages and reduces the time required for transmission. First, here is how a typical message will sound without use of the code.

Station A: "This is KHA3653, unit 1 calling mobile unit 3. Over."
Station B: "This is mobile unit 3, KHA3653. Over."
Station A: "Come back to the office. Over."
Station B: "I can't right now, I'm busy checking a street-light installation. Over."
Station A: "Where are you now? Over."
Station B: "I'm at the corner of 10th and Main. Over."
Station A: "Get through with that job as soon as possible and then come to the office. OK? Over."
Station B: "All right, I am almost through. I'll be there shortly. This is KHA3653, mobile unit 3—over and out."
Station A: "This is KHA3653, unit 1—over and out."

This conversation is typical of thousands transmitted every day on the Citizens band. Now let's see how the same message would sound using the radiotelephone code in Table 8-1.

Station A: "This is KHA3653, unit 1 calling mobile unit 3. Over."
Station B: "This is mobile unit 3, KHA3653. Over."
Station A: "10-19."
Station B: "10-6."
Station A: "10-20."
Station B: "Corner of 10th and Main."
Station A: "10-18, 10-19, 10-4?"
Station B: "10-4. Mobile unit 3, KHA3653—over and out."
Station A: "KHA3653, unit 1—over and out."

Table 8-1. Radiotelephone Code

Code No.	Meaning
10-1	Receiving poorly.
10-2	Receiving well.
10-3	Stand by.
10-4	O.K.
10-5	Relay message.
10-6	Busy at present.
10-7	Out of service.
10-8	In service.
10-9	Repeat.
10-10	Out of service—subject to call.
10-11	Transmitting too rapidly.
10-12	Officials or visitors present.
10-13	Advise weather and road conditions.
10-18	Complete assignment as quickly as possible.
10-19	Come to office.
10-20	Your location?
10-21	Call this station by phone.
10-22	Disregard.
10-30	Does not conform to rules and regulations.
10-33	Emergency traffic at this station.
10-36	Correct time?
10-42	Out of service at home.
10-45	Call ——————— by phone.
10-60	What is next message number?
10-64	Not clear.
10-71	Proceed with transmission in sequence.
10-84	What is your telephone number?
10-92	Your quality poor; transmitter apparently out of adjustment.
10-93	Frequencies to be checked.
10-94	Test intermittently with normal modulation for . . .
10-95	Test with no modulation.
10-99	Unable to receive your signals.

Obviously, the same basic information was transmitted with this method, but in much less time. It is desirable to use code to some extent, especially in a business that requires heavy traffic such as dispatching delivery trucks, taxis, etc.

Improper Operation

Improper station operation can be classified into one of two categories—accidental and willful. Possibly a third category could be added to include those who have not taken the time to learn the rules, or who have been misinformed somewhere along the line. In any event, there is an increasing number of violations occurring in all categories.

The FCC operates a number of monitoring stations twenty-four hours a day. These stations are manned by engineers who, with the help of the most modern and efficient radio receivers and tracking devices, scan the frequency spectrum for violations committed in any of the radio services. In policing the Citizens bands (particularly the Class-D Service) the FCC monitoring stations are uncovering violations by the thousands. This, of course, has resulted in increased regulation of this service and stricter enforcement of Part 95 of the FCC *Rules and Regulations.* These regulations are altered and amended from time to time as the need arises, and it appears that this need is becoming much more frequent. One approach to stemming the increasing number of violations is to impose heavy fines on violators, and this has been done in a number of instances. According to recent reports, violations fall mainly into two groups—off-frequency operation and improper use of CB stations.

Off-frequency operation can be the result of an inexperienced or unqualified person tampering with the transmitter, or the equipment may have become defective or is badly in need of frequency adjustment. How often the frequency of a transmitter should be checked will depend on a number of factors. Some units require frequency checks every few months, whereas others may not require it for a year or more. The best safeguard against off-frequency operation is periodic checks by a qualified person having the proper equipment and skill.

Another common mistaken belief is that Citizens radio can be used in the same manner as ham radio. This has been especially noticeable in the Class-D service. Perhaps the reason is that Class-D frequencies occupy a space formerly allocated to the 11-meter amateur band. Some who desired to become ham operators, but for some reason could not make the grade, have resorted to Citizens radio, feeling that it can be used in the same manner. Occasionally an amateur-radio operator, who feels the loss of the 11-meter band unjust, has entered the CB ranks for the purpose of continuing ham-style operation in this band. Fortunately, such cases are in the minority.

Such practices as calling *CQ* (any station answer) to establish contacts with unknown operators is contrary to proper operating practices. This would be permissible only in an emergency where it was imperative that a unit within the area be contacted. In general operation, however, all calls should be directed to the specific station or stations controlled by the licensee. Contacts with those outside this network should be made only by prearrangement or in an emergency. This is done to discourage activities such as DXing (attempting to communicate with other stations solely for the purpose of achieving maximum distance), to see how many contacts can be made with other stations, or to just plain reduce the temptation for casual visiting and "rag chewing" on the air. DXing ("working the skip") on the CB frequencies is prohibited by the FCC. The regulations state that no CB stations, even under the same license, can communicate with other stations located more than 150 miles away, even if conditions permit. Furthermore, the FCC has not only limited the transmitter input power but the rf power output as well.

Another improper operating practice that has become quite a problem is excessive on-the-air testing. This type of testing should be kept to a minimum because it only adds to the channel congestion.

Citizens-band equipment cannot be used for any purpose contrary to federal, state, and local laws. The meaning of this should be apparent—obviously you cannot use CB equipment to aid in a break-in, bank robbery, or any other crime. Moreover, you cannot use Citizens-band equipment for conveying any type of program material (music, broadcasting, etc.) either directly or indirectly (see Article 95.83(a) of the FCC rules in the Appendix). Also, while it is permissible for the licensee to allow other persons to use the equipment, he cannot charge a fee for this service except as outlined in 95.87.

The licensee, unless possessing a valid first- or second-class Commercial Radiotelephone Operator's License, is not permitted to make any adjustments, repairs, or modifications that could cause improper transmission of rf energy. The foregoing can only be performed by or under the direct supervision of a person holding such a license. Occasionally, an individual will attempt a modification to "soup up" the transmitter. This can lead to nothing but trouble. Furthermore, any proposed changes that affect the terms of the current license require prior approval from the FCC.

Other on-the-air practices that are strictly forbidden include the use of false call letters, and any type of profane, indecent, or obscene language. The FCC monitoring stations are very effective in tracking down persons operating equipment contrary to any of the rules and regulations governing this service, and a violation may result in a written citation. A reply to this citation must be made in the manner prescribed in Article 95.113 of the FCC regulations. A continuance of such operation can result in a fine and/or revocation of the station license and loss of all operating privileges.

Operating Requirements

Citizens radio equipment operated at a fixed location must have a current license (or a photocopy thereof) permanently posted at the key point of control for such operation. Furthermore, a photocopy of this license must be posted at all other fixed locations from which the station is controlled (see Article 95.101 of the FCC regulations in the Appendix).

The serial number on each citizens radio station license serves as the call letters for the station. The complete assigned call letters of the CB station must be given (in English) at the beginning and end of all communications. Only standard nationally or internationally recognized phonetic alphabets can be used for the call letters.

All communications or signals must be restricted to the minimum practical transmission time. In any case, the transmission between different Class-D stations can not exceed 5 minutes during normal operation. At the conclusion of the 5 minute period or at the termination of an exchange lasting for a shorter period, 5 minutes of silence must be observed before any further transmissions can be made. All channels concerned must then be monitored before communications are resumed. Stations used only for radio control (Class C) need not identify their transmissions.

Should normal communications be inadequate or be disrupted during an earthquake, hurricane, flood, or similar disaster, CB equipment in any class can be operated on an emergency basis in compliance with the provisions stated in Article 95.85 governing permissible communications. However, as soon as possible after such emergency operation begins, a notice must be sent to either the FCC in Washington, D.C., or to the Engineer-in-Charge of the nearest field office, stating how the equipment is being used and what type of emergency exists. As soon as normal communicating facilities are restored, such operation must be discontinued and the FCC notified that special operations have ceased. Discontinuance of special operations may be ordered by the FCC at any time.

THE NEED FOR REGULATIONS

The FCC regulations are not designed to limit the usefulness of CB radio, but instead are intended to discourage improper operation and make CB radio serve the purpose for which it is intended—providing short-distance personal and business radiocommunications.

9

Canadian CB Radio

Canada also has its Citizens band radio but here it is known as the General Radio Service. In many ways the Citizens Radio and General Radio Services are similar—both, for example, are designed for either business or personal use.

THE GENERAL RADIO SERVICE

The General Radio Service (GRS) was established in April of 1962 by the Department of Transport in Ottawa. This department is the controlling body of the service and acts in a capacity similar to the FCC in the United States. The radio service itself, although officially recognized at the GRS, is generally referred to as CB radio by most Canadian operators. Since the General Radio Service was initiated nearly four years after the 11-meter frequencies were allocated for CB use in the United States, it gave the Canadian officials time to profit from many of our mistakes.

The number of licensed stations in the GRS is far less than the number of CB stations in the States. Moreover, the Department of Transport required that each transmitter be licensed separately rather than allowing a number of units to be operated under the same license.

License Requirements

To obtain a license in the General Radio Service the applicant must be at least 18 years of age and have a direct need for radiocommunications with stations operating under a similar license. The applicant must be a British subject or a business which is incorporated within the British Commonwealth. License applications can be obtained either by mail or in person from the nearest District Office of the Department of Transport. If the address of the office is not known, it can be obtained by writing the Department of Transportation, Air Services Branch, Hunter Building, Ottawa, Ontario.

Each transmitter must be licensed separately and the applicant is required to pay a fee for each license. The license is valid for a period of three years after the first day of April of the fiscal year in which it is issued, and the license is not transferable.

There are six radio districts in Canada (XM1 through XM6). Each licensing office within these districts is responsible for governing the operation of stations located within its area. This includes the issuing of all notices of violation, the imposing of fines, etc.

Operating Requirements

The rules and regulations governing the General Radio Service are very similar to those for CB in the United States. They provide for emergency communications and prohibit such things as the use of false call letters and distress signals, profane or obscene language, etc. Stations cannot use this radio service for recreation or as a hobby and are not permitted to communicate with stations in other radio services. Moreover, no calls can be directed to another GRS station located beyond the normal ground-wave range. Although a full 5 watts of plate power input is permitted in the final rf transmitter stage, the rf output is limited to 3 watts (1 watt less than that allowed CB stations in the United States). Transmitters operating with 100 milliwatts or less plate power input to the final do not require a license and may communicate with other GRS stations on any of the channels from 26.97 to 27.27 MHz. (Only a short time ago these low-power transmitters were prohibited from using any of the GRS channels.)

The GRS channel frequencies are the same as those for CB in the United States; however, operation is only permitted on 19 of the 23 channels. Channels 1, 2, 3, and 23 are reserved for municipal use only. Canadian operators are not permitted to communicate across the U.S. border with CB operators but CB'ers from the U.S. may obtain a temporary license to operate while visiting or traveling through Canada.

The Tourist Radio Service (TRS)

In 1964 the Department of Transport established what is known as the Tourist Radio Service. This service allows those holding valid Citizens-band authorizations to obtain a temporary license to operate their equipment while visiting or traveling through Canada. There is no fee for this license and it is valid for one year. It may be renewed at the end of that time but it must be pointed out that it is not transferable. Furthermore, the license must be in the possession of the operator at all times. The holder must adhere to all existing GRS regulations and is only allowed to operate on CB channels 4 through 22.

Anyone desiring to obtain a temporary license in the Tourist Radio Service should make application at least 30 days prior to entry and at the nearest port of entry. Requests for applications should be addressed to the Regional Superintendent, Radio Regulations, Department of Transport, in care of the appropriate Regional Office. The inquiry should include your name and address, CB call sign, class of service, and the length of time you intend to be in Canada.

Radio Equipment

The radio equipment employed in the General Radio Service is similar to that used in the States; however, it must conform to all of the technical requirements set forth in Radio Standards Specification 136 issued by the Department of Transport.

APPENDIX I

FCC Rules and Regulations

In filing an application for a Citizens-band station license (Form 505), you must certify that you have a current copy of the FCC *Rules and Regulations,* Part 95. For your convenience, Part 95 has been included in this Appendix. Although the Rules and Regulations reproduced here are current as of September 15, 1975, they are subject to amendment and thus may or may not fulfill the requirements of FCC Form 505 at a later date.

Contents—Part 95

Subpart A—General

Subpart B—Applications and Licenses

Subpart C—Technical Regulations

Subpart D—Station Operating Requirements

Subpart E—Operation of Citizens Radio Stations in the United States by Canadians

AUTHORITY: §§ 95.1 to 95.147 issued under secs. 4, 303, 48 Stat. 1066, 1082, as amended; 47 U.S.C. 154, 303. Interpret or apply 48 Stat. 1064–1068, 1081–1105, as amended; 47 U.S.C. Sub-chap. I, III–VI.

SUBPART A—GENERAL

§ 95.1 Basis and purpose.

The rules and regulations set forth in this part are issued pursuant to the provisions of Title III of the Communications Act of 1934, as amended, which vests authority in the Federal Communications Commission to regulate radio transmissions and to issue licenses for radio stations. These rules are designed to provide for private short-distance radiocommunications service for the business or personal activities of licensees, for radio signaling, for the control of remote objects or devices by means of radio; all to the extent that these uses are not specifically prohibited in this part. They also provide for procedures whereby manufacturers of radio equipment to be used or operated in the Citizens Radio Service may obtain type acceptance and/or type approval of such equipment as may be appropriate.

§ 95.3 Definitions.

For the purpose of this part, the following definitions shall be applicable. For other definitions, refer to Part 2 of this chapter.

(a) Definitions of services.

Citizens Radio Service. A radiocommunications service of fixed, land, and mobile stations intended for short-distance personal or business radiocommunications, radio signaling, and control of remote objects or devices by radio; all to the extent that these uses are not specifically prohibited in this part.

Fixed service. A service of radiocommunication between specified fixed points.

Mobile service. A service of radiocommunication between mobile and land stations or between mobile stations.

(b) Definitions of stations.

Base station. A land station in the land mobile service carrying on a service with land mobile stations.

Class A station. A station in the Citizens Radio Service licensed to be operated on an assigned frequency in the 460–470 MHz band with a transmitter output power of not more than 50 watts.

Class B station. (All operations terminated as of November 1, 1971.)

Class C station. A station in the Citizens Radio Service licensed to be operated on an authorized frequency in the 26.96–27.23 MHz band, or on the frequency 27.255 MHz, for the control of remote objects or devices by radio, or for the remote actuation of devices which are used solely as a means of attracting attention, or on an authorized frequency in the 72–76 MHz band for the radio control of models used for hobby purposes only.

Class D station. A station in the Citizens Radio Service licensed to be operated for radiotelephony, only, on an authorized frequency in the 26.96–27.23 MHz band and on the frequency 27.255 MHz.

Fixed station. A station in the fixed service.

Land station. A station in the mobile service not intended for operation while in motion. (Of the various types of land stations, only the base station is pertinent to this part.)

Mobile station. A station in the mobile service intended to be used while in motion or during halts at unspecified points. (For the purposes of this part, the term includes hand-carried and pack-carried units.)

(c) Miscellaneous definitions.

Antenna structures. The term "antenna structures" includes the radiating system, its supporting structures and any appurtenances mounted thereon.

Assigned frequency. The frequency appearing on a station authorization from which the carrier frequency may deviate by an amount not to exceed that permitted by the frequency tolerance.

Authorized bandwidth. The maximum permissible bandwidth for the particular emission used. This shall be the occupied bandwidth or necessary bandwidth, whichever is greater.

Carrier power. The average power at the output terminals of a transmitter (other than a transmitter having a suppressed, reduced or controlled carrier) during one radio frequency cycle under conditions of no modulation.

Control point. A control point is an operating position which is under the control and supervision of the licensee, at which a person immediately responsible for the proper operation of the transmitter is stationed, and at which adequate means are available to aurally monitor all transmissions and to render the transmitter inoperative.

Dispatch point. A dispatch point is any position from which messages may be transmitted under the supervision of the person at a control point.

Double sideband emission. An emission in which both upper and lower sidebands resulting from the modulation of a particular carrier are transmitted. The carrier, or a portion thereof, also may be present in the emission.

External radio frequency power amplifiers. As defined in § 2.815(a) and as used in this part, an external radio frequency power amplifier is any device which, (1) when used in conjunction with a radio transmitter as a signal source is capable of amplification of that signal, and (2) is not an integral part of a radio transmitter as manufactured.

Harmful interference. Any emission, radiation or induction which endangers the functioning of a radionavigation service or other safety service or seriously degrades, obstructs or repeatedly interrupts a radiocommunication service operating in accordance with applicable laws, treaties, and regulations.

Man-made structure. Any construction other than a tower, mast or pole.

Mean power. The power at the output terminals of a transmitter during normal operation, averaged over a time sufficiently long compared with the period of the lowest frequency encountered in the modulation. A time of $1/10$ second during which the mean power is greatest will be selected normally.

Necessary bandwidth. For a given class of emission, the minimum value of the occupied bandwidth sufficient to ensure the transmission of information at the rate and with the quality required for the system employed, under specified conditions. Emissions useful for the good functioning of the receiving equipment, as for example, the emission corresponding to the carrier of reduced carrier systems, shall be included in the necessary bandwidth.

Occupied bandwidth. The frequency bandwidth such that, below its lower and above its upper frequency limits, the mean powers radiated are each equal to 0.5% of the total mean power radiated by a given emission.

Omnidirectional antenna. An antenna designed so the maximum radiation in any horizontal direction is within 3 dB of the minimum radiation in any horizontal direction.

Peak envelope power. The average power at the output terminals of a transmitter during one radio frequency cycle at the highest crest of the modulation envelope, taken under conditions of normal operation.

Person. The term "person" includes an individual, partnership, association, joint-stock company, trust or corporation.

Remote control. The term "remote control" when applied to the use or operation of a citizens radio station means control of the transmitting equipment of that station from any place other than the location of the transmitting equipment, except that direct mechanical control or direct electrical control by wired connections of transmitting equipment from some other point on the same premises, craft or vehicle shall not be considered to be remote control.

Single sideband emission. An emission in which only one sideband is transmitted. The carrier, or a portion thereof, also may be present in the emission.

Station authorization. Any construction permit, license, or special temporary authorization issued by the Commission.

§ 95.5 Policy governing the assignment of frequencies.

(a) The frequencies which may be assigned to Class A stations in the Citizens Radio Service, and the frequencies which are available for use by Class C or Class D stations are listed in Subpart C of this part. Each frequency available for assignment to, or use by, stations in this service is available on a shared basis only, and will not be assigned for the exclusive use of any one applicant; however, the use of a particular frequency may be restricted to (or in) one or more specified geographical areas.

(b) In no case will more than one frequency be assigned to Class A stations for the use of a single applicant in any given area until it has been demonstrated conclusively to the Commission that the assignment of an additional frequency is essential to the operation proposed.

(c) All applicants and licensees in this service shall cooperate in the selection and use of the frequencies assigned or authorized, in order to minimize interference and thereby obtain the most effective use of the authorized facilities.

(d) Simultaneous operation on more than one frequency in the 72–76 MHz band by a transmitter or transmitters of a single licensee is prohibited whenever such operation will cause harmful interference to the operation of other licensees in this service.

§ 95.6 Types of operation authorized.

(a) Class A stations may be authorized as mobile stations, as base stations, as fixed stations, or as base or fixed stations to be operated at unspecified or temporary locations.

(b) Class C and Class D stations are authorized as mobile stations only; however, they may be operated at fixed locations in accordance with other provisions of this part.

§ 95.7 General citizenship restrictions.

A station license may not be granted to or held by:

(a) Any alien or the representative of any alien;

(b) Any foreign government or the representative thereof;

(c) Any corporation organized under the laws of any foreign government;

(d) Any corporation of which any officer or director is an alien;

(e) Any corporation of which more than one-fifth of the capital stock is owned of record or voted by: Aliens or their representatives; a foreign government or representative thereof; or any corporation organized under the laws of a foreign country;

(f) Any corporation directly or indirectly controlled by any other corporation of which any officer or more than one-fourth of the directors are aliens, if the Commission finds that the public interest will be served by the refusal or revocation of such license; or

(g) Any corporation directly or indirectly controlled by any other corporation of which more than one-fourth of the capital stock is owned of record or voted by: Aliens or their representatives; a foreign government or representatives thereof; or any corporation organized under the laws of a foreign government, if the Commission finds that the public interest will be served by the refusal or revocation of such license.

SUBPART B—APPLICATIONS AND LICENSES

§ 95.11 Station authorization required.

No radio station shall be operated in the Citizens Radio Service except under and in accordance with an authorization granted by the Federal Communications Commission.

§ 95.13 Eligibility for station license.

(a) Subject to the general restrictions of § 95.7, any person is eligible to hold an authorization to operate a station in the Citizens Radio Service: *Provided*, That if an applicant for a Class A or Class D station authorization is an individual or partnership, such individual or each partner is eighteen or more years of age; or if an applicant for a Class C station authorization is an individual or partnership, such individual or each partner is twelve or more years of age. An unincorporated association, when licensed under the provisions of this paragraph, may upon specific prior approval of the Commission provide radiocommunications for its members.

NOTE: While the basis of eligibility in this service includes any state, territorial, or local governmental entity, or any agency operating by the authority of such governmental entity, including any duly authorized state, territorial, or local civil defense agency, it should be noted that the frequencies available to stations in this service are shared without distinction between all licensees and that no protection is afforded to the communications of any station in this service from interference which may be caused by the authorized operation of other licensed stations.

(b) [Reserved]

(c) No person shall hold more than one Class C and one Class D station license.

§ 95.15 Filing of applications.

(a) To assure that necessary information is supplied in a consistent manner by all persons, standard forms are prescribed for use in connection with the majority of applications and reports submitted for Commission consideration. Standard numbered forms applicable to the Citizens Radio Service are discussed in § 95.19 and may be obtained from the Washington, D.C., 20554, office of the Commission, or from any of its engineering field offices.

(b) All formal applications for Class C or Class D new, modified, or renewal station authorizations shall be submitted to the Commission's office at 334 York Street, Gettysburg, Pa. 17325. Applications for Class A station authorizations, applications for consent to transfer of control of a corporation holding any citizens radio station authorization, requests for special temporary authority or other special requests, and correspondence relating to an application for any class citizens radio station authorization shall be submitted to the Commission's Office at Washington, D.C. 20554, and should be directed to the attention of the Secretary. Beginning January 1, 1973, applicants for Class A stations in the Chicago Regional Area, defined in § 95.19, shall submit their applications to the Commission's Chicago Regional Office. The address of the Regional Office will be announced at a later date. Applications involving Class A or Class D station equipment which is neither type approved nor crystal controlled, whether of commercial or home construction, shall be accompanied by supplemental data describing in detail the design and construction of the transmitter and methods employed in testing it to determine compliance with the technical requirements set forth in Subpart C of this part.

(c) Unless otherwise specified, an application shall be filed at least 60 days prior to the date on which it is desired that Commission action thereon be completed. In any case where the applicant has made timely and sufficient application for renewal of license, in accordance with the Commission's rules, no license with reference to any activity of a continuing nature shall expire until such application shall have been finally determined.

(d) Failure on the part of the applicant to provide all the information required by the application form, or to supply the necessary exhibits or supplementary statements may constitute a defect in the application.

(e) Applicants proposing to construct a radio station on a site located on land under the jurisdiction of the U.S. Forest Service, U.S. Department of Agriculture, or the Bureau of Land Management, U.S. Department of the Interior, must supply the information and must follow the procedure prescribed by § 1.70 of this chapter.

§ 95.17 Who may sign applications.

(a) Except as provided in paragraph (b) of this section, applications, amendments thereto, and related statements of fact required by the Commission shall be personally signed by the applicant, if the applicant is an individual; by one of the partners, if the applicant is a partnership; by an officer, if the applicant is a corporation; or by a member who is an officer, if the applicant is an unincorporated association. Applications, amendments, and related statements of fact filed on behalf of eligible government entities, such as states and territories of the United States and political subdivisions thereof, the District of Columbia, and units of local government, including incorporated municipalities, shall be signed by such duly elected or appointed officials as may be competent to do so under the laws of the applicable jurisdiction.

(b) Applications, amendments thereto, and related statements of fact required by the Commission may be signed by the applicant's attorney in case of the applicant's physical disability or of his absence from the United States. The attorney shall in that event separately set forth the reason why the application is not signed by the applicant. In addition, if any matter is stated on the basis of the attorney's belief only (rather than his knowledge), he shall separately set forth his reasons for believing that such statements are true.

(c) Only the original of applications, amendments, or related statements of fact need be signed; copies may be conformed.

(d) Applications, amendments, and related statements of fact need not be signed under oath. Willful false statements made therein, however, are punishable by fine and imprisonment, U.S. Code, Title 18, section 1001, and by appropriate administrative sanctions, including revocation of station license pursuant to section 312(a)(1) of the Communications Act of 1934, as amended.

§ 95.19 Standard forms to be used.

(a) *FCC Form 505, Application for Class C or D Station License in the Citizens Radio Service.* This form shall be used when:

(1) Application is made for a new Class C or Class D authorization. A separate application shall be submitted for each proposed class of station.

(2) Application is made for modification of any existing Class C or Class D station authorization in those cases where prior Commission approval of certain changes is required (see § 95.35).

(3) Application is made for renewal of an existing Class C or Class D station authorization, or for reinstatement of such an expired authorization.

(b) *FCC Form 400, Application for Radio Station Authorization in the Safety and Special Radio Services.* Except as provided in paragraph (d) of this section, this form shall be used when:

(1) Application is made for a new Class A base station or fixed station authorization. Separate applications shall be submitted for each proposed base or fixed station at different fixed locations; however, all equipment intended to be operated at a single fixed location is considered to be one station which may, if necessary, be classed as both a base station and a fixed station.

(2) Application is made for a new Class A station authorization for any required number of mobile units (including hand-carried and pack-carried units) to be operated as a group in a single radiocommunication system in a particular area. An application for Class A mobile station authorization may be combined with the application for a single Class A base station authorization when such mobile units are to be operated with that base station only.

(3) Application is made for station license of any Class A base station or fixed station upon completion of construction or installation in accordance with the terms and conditions set forth in any construction permit required to be issued for that station, or application for extension of time within which to construct such a station.

(4) Application is made for modification of any existing Class A station authorization in those cases where prior Commission approval of certain changes is required (see § 95.35).

(5) Application is made for renewal of an existing Class A station authorization, or for reinstatement of such an expired authorization.

(6) Each applicant in the Safety and Special Radio Services (1) for modification of a station license involving a site change or a substantial increase in tower height or (2) for a license for a new station must, before commencing construction, supply the environmental information, where required, and must follow the procedure prescribed by Subpart I of Part 1 of this chapter (§§ 1.1301 through 1.1319) unless Commission action authorizing such construction would be a minor action with the meaning of Subpart I of Part 1.

(7) Application is made for an authorization for a new Class A base or fixed station to be operated at unspecified or temporary locations. When one or more individual transmitters are each intended to be operated as a base station or as a fixed station at unspecified or temporary locations for indeterminate periods, such transmitters may be considered to comprise a single station intended to be operated at temporary locations. The application shall specify the general geographic area within which the operation will be confined. Sufficient data must be submitted to show the need for the proposed area of operation.

(c) *FCC Form 703, Application for Consent to Transfer of Control of Corporation Holding Construction Permit or Station License.* This form shall be used when application is made for consent to transfer control of a corporation holding any citizens radio station authorization.

(d) Beginning April 1, 1972, FCC Form 425 shall be used in lieu of FCC Form 400, applicants for Class A stations located in the Chicago Regional Area defined to consist of the counties listed below:

ILLINOIS

1. Boone.
2. Bureau.
3. Carroll.
4. Champaign.
5. Christian.
6. Clark.
7. Coles.
8. Cook.
9. Cumberland.
10. De Kalb.
11. De Witt.
12. Douglas.
13. Du Page.
14. Edgar.
15. Ford.
16. Fulton.
17. Grundy.
18. Henry.
19. Iroquois.
20. Jo Daviess.
21. Kane.
22. Kankakee.
23. Kendall.
24. Knox.
25. Lake.
26. La Salle.
27. Lee.
28. Livingston.
29. Logan.
30. Macon.
31. Marshall.
32. Mason.
33. McHenry.
34. McLean.
35. Menard.
36. Mercer.
37. Moultrie.
38. Ogle.
39. Peoria.
40. Piatt.
41. Putnam.
42. Rock Island.
43. Sangamon.
44. Shelby.
45. Stark.
46. Stephenson.
47. Tazewell.
48. Vermilion.
49. Warren.
50. Whiteside.
51. Will.
52. Winnebago.
53. Woodford.

INDIANA

1. Adams.
2. Allen.
3. Benton.
4. Blackford.
5. Boone.
6. Carroll.
7. Cass.
8. Clay.
9. Clinton.
10. De Kalb.
11. Delaware.
12. Elkhart.
13. Fountain.
14. Fulton.
15. Grant.
16. Hamilton.
17. Hancock.
18. Hendricks.
19. Henry.
20. Howard.
21. Huntington.
22. Jasper.
23. Jay.
24. Kosciusko.
25. Lake.
26. Lagrange.
27. La Porte.
28. Madison.
29. Marion.
30. Marshall.

INDIANA—Continued

31. Miami.	43. Starke.
32. Montgomery.	44. Steuben.
33. Morgan.	45. Tippecanoe.
34. Newton.	46. Tipton.
35. Noble.	47. Vermilion.
36. Owen.	48. Vigo.
37. Parke.	49. Wabash.
38. Porter.	50. Warren.
39. Pulaski.	51. Wells.
40. Putnam.	52. White.
41. Randolph.	53. Whitley.
42. St. Joseph.	

IOWA

1. Cedar.	5. Jones.
2. Clinton.	6. Muscatine.
3. Dubuque.	7. Scott.
4. Jackson.	

MICHIGAN

1. Allegan.	13. Kalamazoo.
2. Barry.	14. Kent.
3. Berrien.	15. Lake.
4. Branch.	16. Mason.
5. Calhoun.	17. Mecosta.
6. Cass.	18. Montcalm.
7. Clinton.	19. Muskegon.
8. Eaton.	20. Newaygo.
9. Hillsdale.	21. Oceana.
10. Ingham.	22. Ottawa.
11. Ionia.	23. St. Joseph.
12. Jackson.	24. Van Buren.

OHIO

1. Defiance.	4. Van Wert.
2. Mercer.	5. Williams.
3. Paulding.	

WISCONSIN

1. Adams.	18. Manitowoc.
2. Brown.	19. Marquette.
3. Calumet.	20. Milwaukee.
4. Columbia.	21. Outagamie.
5. Dane.	22. Ozaukee.
6. Dodge.	23. Racine.
7. Door.	24. Richland.
8. Fond du Lac.	25. Rock.
9. Grant.	26. Sauk.
10. Green.	27. Sheboygan.
11. Green Lake.	28. Walworth.
12. Iowa.	29. Washington.
13. Jefferson.	30. Waukesha.
14. Juneau.	31. Waupaca.
15. Kenosha.	32. Waushara.
16. Kewaunee.	33. Winnebago.
17. Lafayette.	

§ 95.25 Amendment or dismissal of application.

(a) Any application may be amended upon request of the applicant as a matter of right prior to the time the application is granted or designated for hearing. Each amendment to an application shall be signed and submitted in the same manner and with the same number of copies as required for the original application.

(b) Any application may, upon written request signed by the applicant or his attorney, be dismissed without prejudice as a matter of right prior to the time the application is granted or designated for hearing.

§ 95.27 Transfer of license prohibited.

A station authorization in the Citizens Radio Service may not be transferred or assigned. In lieu of such transfer or assignment, an application for new station authorization shall be filed in each case, and the previous authorization shall be forwarded to the Commission for cancellation.

§ 95.29 Defective applications.

(a) If an applicant is requested by the Commission to file any documents or information not included in the prescribed application form, a failure to comply with such request will constitute a defect in the application.

(b) When an application is considered to be incomplete or defective, such application will be returned to the applicant, unless the Commission may otherwise direct. The reason for return of the applications will be indicated, and if appropriate, necessary additions or corrections will be suggested.

§ 95.31 Partial grant.

Where the Commission, without a hearing, grants an application in part, or with any privileges, terms, or conditions other than those requested, the action of the Commission shall be considered as a grant of such application unless the applicant shall, within 30 days from the date on which such grant is made, or from its effective date if a later date is specified, file with the Commission a written rejection of the grant as made. Upon receipt of such rejection, the Commission will vacate its original action upon the application and, if appropriate, set the application for hearing.

§ 95.33 License term.

Licenses for stations in the Citizens Radio Service will normally be issued for a term of 5 years from the date of original issuance, major modification, or renewal.

§ 95.35 Changes in transmitters and authorized stations.

Authority for certain changes in transmitters and authorized stations must be obtained from the Commission before the changes are made, while other changes do not require prior Commission approval. The following paragraphs of this section describe the conditions under which prior Commission approval is or is not necessary.

(a) Proposed changes which will result in operation inconsistent with any of the terms of the current authorization require that an application for modification of license be submitted to the Commission. Application for modification shall be submitted in the same manner as an application for a new station license, and the licensee shall forward his existing authorization to the Commission for cancellation immediately upon receipt of the superseding authorization. Any of the following changes to authorized stations may be made only upon approval by the Commission:

(1) Increase the overall number of transmitters authorized.

(2) Change the presently authorized location of a Class A fixed or base station or control point.

(3) Move, change the height of, or erect a Class A station antenna structure.

(4) Make any change in the type of emission or any increase in bandwidth of emission or power of a Class A station.

(5) Addition or deletion of control point(s) for an authorized transmitter of a Class A station.

(6) Change or increase the area of operation of a Class A mobile station or a Class A base or fixed station authorized to be operated at temporary locations.

(7) Change the operating frequency of a Class A station.

(b) When the name of a licensee is changed (without changes in the ownership, control, or corporate structure), or when the mailing address of the licensee is changed (without changing the authorized location of the base or fixed Class A station) a formal application for modification of the license is not required. However, the licensee shall notify the Commission promptly of these changes. The notice, which may be in letter form, shall contain the name and address of the licensee as they appear in the Commission's records, the new name and/or address, as the case may be, and the call signs and classes of all radio stations authorized to the licensee under this part. The notice concerning Class C or D radio stations shall be sent to Federal Communications Commission, Gettysburg, Pa. 17325, and a copy shall be maintained with the records of the station. The notice concerning Class A stations shall be sent to (1) Secretary, Federal Communications Commission, Washington, D.C. 20554, and (2) to Engineer in Charge of the Radio District in which the station is located, and a copy shall be maintained with the license of the station until a new license is issued.

(c) Proposed changes which will not depart from any of the terms of the outstanding authorization for the station may be made without prior Commission approval. Included in such changes is the substitution of transmitting equipment at any station, provided that the equipment employed is included in the Commission's "Radio Equipment List," and is listed as acceptable for use in the appropriate class of station in this service. Provided it is crystal-controlled and otherwise complies with the power, frequency tolerance, emission and modulation percentage limitations prescribed, non-type accepted equipment may be substituted at:

(1) Class C stations operated on frequencies in the 26.99–27.26 MHz band;

(2) Class D stations until November 22, 1974.

(d) Transmitting equipment type accepted for use in Class D stations shall not be modified by the user. Changes which are specifically prohibited include:

(1) Internal or external connection or addition of any part, device or accessory not included by the manufacturer with the transmitter for its type acceptance. This shall not prohibit the external connection of antennas or antenna transmission lines, antenna switches, passive networks for coupling transmission lines or antennas to transmitters, or replacement of microphones.

(2) Modification in any way not specified by the transmitter manufacturer and not approved by the Commission.

(3) Replacement of any transmitter part by a part having different electrical characteristics and ratings from that replaced unless such part is specified as a replacement by the transmitter manufacturer.

(4) Substitution or addition of any transmitter oscillator crystal unless the crystal manufacturer or transmitter manufacturer has made an express determination that the crystal type, as installed in the specific transmitter type, will provide that transmitter type with the capability of operating within the frequency tolerance specified in Section 95.45(a).

(5) Addition or substitution of any component, crystal or combination of crystals, or any other alteration to enable transmission on any frequency not authorized for use by the licensee.

(e) Only the manufacturer of the particular unit of equipment type accepted for use in Class D stations may make the permissive changes allowed under the provisions of Part 2 of this chapter for type acceptance. However, the manufacturer shall not make any of the following changes to the transmitter without prior written authorization from the Commission:

(1) Addition of any accessory or device not specified in the application for type acceptance and approved by the Commission in granting said type acceptance.

(2) Addition of any switch, control, or external connection.

(3) Modification to provide capability for an additional number of transmitting frequencies.

§ 95.37 Limitations on antenna structures.

(a) Except as provided in paragraph (b) of this section, an antenna for a Class A station which exceeds the following height limitations may not be erected or used unless notice has been filed with both the FAA on FAA Form 7460–1 and with the Commission on Form 714 or on the license application form, and prior approval by the Commission has been obtained for:

(1) Any construction or alteration of more than 200 feet in height above ground level at its site (§ 17.7(a) of this chapter).

(2) Any construction or alteration of greater height than an imaginary surface extending outward and upward at one of the following slopes (§ 17.7(b) of this chapter):

(i) 100 to 1 for a horizontal distance of 20,000 feet from the nearest point of the nearest runway of each airport with at least one runway more than 3,200 feet in length, excluding heliports, and seaplane bases without specified boundaries, if that airport is either listed in the Airport Directory of the current Airman's Information Manual or is operated by a Federal military agency.

(ii) 50 to 1 for a horizontal distance of 10,000 feet from the nearest point of the nearest runway of each airport with its longest runway no more than 3,200 feet in length, excluding heliports, and seaplane bases without specified boundaries, if that airport is either listed in the Airport Directory or is operated by a Federal military agency.

(iii) 25 to 1 for a horizontal distance of 5,000 feet from the nearest point of the nearest landing and takeoff area of each heliport listed in the Airport Directory or operated by a Federal military agency.

(3) Any construction or alteration on any airport listed in the Airport Directory of the current Airman's Information Manual (§ 17.7(c) of this chapter).

(b) A notification to the Federal Aviation Administration is not required for any of the following construction or alteration of Class A station antenna structures.

(1) Any object that would be shielded by existing structures of a permanent and substantial character or by natural terrain or topographic features of equal or greater height, and would be located in the congested area of a city, town, or settlement where it is evident beyond all reasonable doubt that the structure so shielded will not adversely affect safety in air navigation. Applicants claiming such exemption shall submit a statement with their application to the Commission explaining the basis in detail for their finding (§ 17.14(a) of this chapter).

(2) Any antenna structure of 20 feet or less in height except one that would increase the height of another antenna structure (§17.14(b) of this chapter).

(c) A Class C or Class D station operated at a fixed location shall employ a transmitting antenna which complies with at least one of the following:

(1) The antenna and its supporting structure does not exceed 20 feet in height above ground level; or

(2) The antenna and its supporting structure does not exceed by more than 20 feet the height of any natural formation, tree or man-made structure on which it is mounted; or

NOTE: A man-made structure is any construction other than a tower, mast, or pole.

(3) The antenna is mounted on the transmitting antenna structure of another authorized radio station and does not exceed the height of the antenna supporting structure of the other station; or

(4) The antenna is mounted on and does not exceed the height of the antenna structure otherwise used solely for receiving purposes, which structure itself complies with subparagraph (1) or (2) of this paragraph.

(5) The antenna is omnidirectional and the highest point of the antenna and its supporting structure does not exceed 60 feet above ground level and the highest point also does not exceed one foot in height above the established airport elevation for each 100 feet of horizontal distance from the nearest point of the nearest airport runway.

NOTE: A work sheet will be made available upon request to assist in determining the maximum permissible height of an antenna structure.

(d) Class C stations operated on frequencies in the 72–76 MHz band shall employ a transmitting antenna which complies with all of the following:

(1) The gain of the antenna shall not exceed that of a half-wave dipole;

(2) The antenna shall be immediately attached to, and an integral part of, the transmitter; and

(3) Only vertical polarization shall be used.

(e) Further details as to whether an aeronautical study and/or obstruction marking and lighting may be required, and specifications for obstruction marking and lighting when required, may be obtained from Part 17 of this chapter, "Construction, Marking, and Lighting of Antenna Structures."

(f) Subpart I of Part 1 of this chapter contains procedures implementing the National Environmental Policy Act of 1969. Applications for authorization of the construction of certain classes of communications facilities defined as "major actions" in § 1.305 thereof, are required to be accompanied by specified statements. Generally these classes are:

(1) Antenna towers or supporting structures which exceed 300 feet in height and are not located in areas devoted to heavy industry or to agriculture.

(2) Communications facilities to be located in the following areas:

(i) Facilities which are to be located in an officially designated wilderness area or in an area whose designation as a wilderness is pending consideration;

(ii) Facilities which are to be located in an officially designated wildlife preserve or in an area whose designation as a wildlife preserve is pending consideration;

(iii) Facilities which will affect districts, sites, buildings, structures or objects, significant in American history, architecture, archaeology or culture, which are listed in the National Register of Historic Places or are eligible for listing (see 36 CFR 800.2 (d) and (f) and 800.10); and

(iv) Facilities to be located in areas which are recognized either nationally or locally for their special scenic or recreational value.

(3) Facilities whose construction will involve extensive change in surface features (e.g. wetland fill, deforestation or water diversion).

NOTE: The provisions of this paragraph do not include the mounting of FM, television or other antennas comparable thereto in size on an existing building or antenna tower. The use of existing routes, buildings and towers is an environmentally desirable alternative to the construction of new routes or towers and is encouraged.

If the required statements do not accompany the application, the pertinent facts may be brought to the attention of the Commission by any interested person during the course of the license term and considered de novo by the Commission.

SUBPART C—TECHNICAL REGULATIONS

§ 95.41 Frequencies available.

(a) Frequencies available for assignment to Class A stations:

(1) The following frequencies or frequency pairs are available primarily for assignment to base and

mobile stations. They may also be assigned to fixed stations as follows:

(i) Fixed stations which are used to control base stations of a system may be assigned the frequency assigned to the mobile units associated with the base station. Such fixed stations shall comply with the following requirements if they are located within 75 miles of the center of urbanized areas of 200,000 or more population.

(*a*) If the station is used to control one or more base stations located within 45 degrees of azimuth, a directional antenna having a front-to-back ratio of at least 15 dB shall be used at the fixed station. For other situations where such a directional antenna cannot be used, a cardioid, bidirectional or omnidirectional antenna may be employed. Consistent with reasonable design, the antenna used must, in each case, produce a radiation pattern that provides only the coverage necessary to permit satisfactory control of each base station and limit radiation in other directions to the extent feasible.

(*b*) The strength of the signal of a fixed station controlling a single base station may not exceed the signal strength produced at the antenna terminal of the base receiver by a unit of the associated mobile station, by more than 6 dB. When the station controls more than one base station, the 6 dB control-to-mobile signal difference need be verified at only one of the base station sites. The measurement of the signal strength of the mobile unit must be made when such unit is transmitting from the control station location or, if that is not practical, from a location within one-fourth mile of the control station site.

(*c*) Each application for a control station to be authorized under the provisions of this paragraph shall be accompanied by a statement certifying that the output power of the proposed station transmitter will be adjusted to comply with the foregoing signal level limitation. Records of the measurements used to determine the signal ratio shall be kept with the station records and shall be made available for inspection by Commission personnel upon request.

(*d*) Urbanized areas of 200,000 or more population are defined in the U.S. Census of Population, 1960, Vol. 1, table 23, page 50. The centers of urbanized areas are determined from the Appendix, page 226 of the U.S. Commerce publication "Air Line Distance Between Cities in the United States."

(ii) Fixed stations, other than those used to control base stations, which are located 75 or more miles from the center of an urbanized area of 200,000 or more population. The centers of urbanized areas of 200,000 or more population are listed on page 226 of the Appendix to the U.S. Department of Commerce publication "Air Line Distance Between Cities in the United States." When the fixed station is located 100 miles or less from the center of such an urbanized area, the power output may not exceed 15 watts. All fixed systems are limited to a maximum of two frequencies and must employ directional antennas with a front-to-back ratio of at least 15 dB. For two-frequency systems, separation between transmit-receive frequencies is 5 MHz.

Base and Mobile (MHz)	*Mobile Only (MHz)*
462.550	467.550
462.575	467.575
462.600	467.600
462.625	467.625
462.650	467.650
462.675	467.675
462.700	467.700
462.725	467.725

(2) Conditions governing the operation of stations authorized prior to March 18, 1968:

(i) All base and mobile stations authorized to operate on frequencies other than those listed in subparagraph (1) of this paragraph may continue to operate on those frequencies only until January 1, 1970.

(ii) Fixed stations located 100 or more miles from the center of any urbanized area of 200,000 or more population authorized to operate on frequencies other than those listed in subparagraph (1) of this paragraph will not have to change frequencies provided no interference is caused to the operation of stations in the land mobile service.

(iii) Fixed stations, other than those used to control base stations, located less than 100 miles (75 miles if the transmitter power output does not exceed 15 watts) from the center of any urbanized area of 200,000 or more population must discontinue operation by November 1, 1971. However, any operation after January 1, 1970, must be on frequencies listed in subparagraph (1) of this paragraph.

(iv) Fixed stations, located less than 100 miles from the center of any urbanized area of 200,000 or more population, which are used to control base stations and are authorized to operate on frequencies other than those listed in subparagraph (1) of this paragraph may continue to operate on those frequencies only until January 1, 1970.

(v) All fixed stations must comply with the applicable technical requirements of subparagraph (1) relating to antennas and radiated signal strength of this paragraph by November 1, 1971.

(vi) Notwithstanding the provisions of subdivisions (i) through (v) of this subparagraph, all stations authorized to operate on frequencies between 465.000 and 465.500 MHz and located within 75 miles of the center of the 20 largest urbanized areas of the United States, may continue to operate on these frequencies only until January 1, 1969. An extension to continue operation on such frequencies until January 1, 1970, may be granted to such station licensees on a case by case basis if the Commission finds that continued operation would not be inconsistent with planned usage of the particular frequency for police purposes. The 20 largest urbanized areas can be found in the U.S. Census of Population, 1960, vol. 1, table 23, page 50. The centers of urbanized areas are determined from the appendix, page 226, of the U.S. Commerce publication, "Air Line Distance Between Cities in the United States."

(b) [Reserved]

(c) Class C mobile stations may employ only amplitude tone modulation or on-off keying of the unmodulated carrier, on a shared basis with other stations in the Citizens Radio Service on the frequencies and under the conditions specified in the following tables:

(1) For the control of remote objects or devices by radio, or for the remote actuation of devices which are used solely as a means of attracting attention and subject to no protection from interference due to the operation of industrial, scientific, or medical devices within the 26.96–27.28 MHz band, the following frequencies are available:

(MHz)	(MHz)	(MHz)
26.995	27.095	27.195
27.045	27.145	[1] 27.255

[1] The frequency 27.255 MHz also is shared with stations in other services.

(2) Subject to the conditions that interference will not be caused to the remote control of industrial equipment operating on the same or adjacent frequencies and to the reception of television transmissions on Channels 4 or 5; and that no protection will be afforded from interference due to the operation of fixed and mobile stations in other services assigned to the same or adjacent frequencies in the band, the following frequencies are available solely for the radio remote control of models used for hobby purposes:

(i) For the radio remote control of any model used for hobby purposes:

MHz	MHz	MHz
72.16	72.32	72.96

(ii) For the radio remote control of aircraft models only:

MHz	MHz	MHz
72.08	72.24	72.40
75.64		

(d) The frequencies listed in the following tables are available for use by Class D mobile stations employing radiotelephony only, on a shared basis with other stations in the Citizens Radio Service, and subject to no protection from interference due to the operation of industrial, scientific, or medical devices within the 26.96–27.28 MHz band.

(1) The following frequencies, commonly known as Channels 1 through 8 and 10 through 23, may be used for communications between units of the same station:

MHz	Channel	MHz	Channel
26.965	1	27.115	13
26.975	2	27.125	14
26.985	3	27.135	15
27.005	4	27.155	16
27.015	5	27.165	17
27.025	6	27.175	18
27.035	7	27.185	19
27.055	8	27.205	20
27.075	10	27.215	21
27.085	11	27.225	22
27.105	12	27.255	23

(2) Only the following frequencies may be used for communications between units of different stations:

MHz	Channel	MHz	Channel
27.075	10	27.125	14
27.085	11	27.135	15
27.105	12	27.255	23
27.115	13		

(3) The frequency 27.065 MHz (Channel 9) shall be used solely for:

(i) Emergency communications involving the immediate safety of life of individuals or the immediate protection of property or

(ii) Communications necessary to render assistance to a motorist.

NOTE: A licensee, before using Channel 9, must make a determination that his communication is either or both (a) an emergency communication or (b) is necessary to render assistance to a motorist. To be an emergency communication, the message must have some direct relation to the immediate safety of life or immediate protection of property. If no immediate action is required, it is not an emergency. What may not be an emergency under one set of circumstances may be an emergency under different circumstances. There are many worthwhile public service communications that do not qualify as emergency communications. In the case of motorist assistance, the message must be necessary to assist a particular motorist and not, except in a valid emergency, motorists in general. If the communications are to be lengthy, the exchange should be shifted to another channel, if feasible, after contact is established. No nonemergency or nonmotorist assistance communications are permitted on Channel 9 even for the limited purpose of calling a licensee monitoring a channel to ask him to switch to another channel. Although Channel 9 may be used for marine emergencies, it should not be considered a substitute for the authorized marine distress system. The Coast Guard has stated it will not "participate directly in the Citizens Radio Service by fitting with and/or providing a watch on any Citizens Band Channel. (Coast Guard Commandant Instruction 2302.6.)"

The following are examples of permitted and prohibited types of communications. They are guidelines and are not intended to be all inclusive.

Permitted	Example message
Yes	"A tornado sighted six miles north of town."
No	"This is observation post number 10. No tornados sighted."
Yes	"I am out of gas on Interstate 95."
No	"I am out of gas in my driveway."
Yes	"There is a four-car collision at Exit 10 on the Beltway, send police and ambulance."
No	"Traffic is moving smoothly on the Beltway."
Yes	"Base to Unit 1, the Weather Bureau has just issued a thunderstorm warning. Bring the sailboat into port."
No	"Attention all motorists. The Weather Bureau advises that the snow tomorrow will accumulate 4 to 6 inches."
Yes	"There is a fire in the building on the corner of 6th and Main Streets."
No	"This is Halloween patrol unit number 3. Everything is quiet here."

The following priorities should be observed in the use of Channel 9.

1. Communications relating to an existing situation dangerous to life or property, i.e., fire, automobile accident.
2. Communications relating to a potentially hazardous situation, i.e., car stalled in a dangerous place, lost child, boat out of gas.
3. Road assistance to a disabled vehicle on the highway or street.
4. Road and street directions.

(e) Upon specific request accompanying application for renewal of station authorization, a Class A station in this service, which was authorized to operate on a frequency in the 460–461 MHz band until March 31, 1967, may be assigned that frequency for continued use until not later than March 31, 1968, subject to all other provisions of this part.

§ 95.43 Transmitter power.

(a) Transmitter power is the power at the transmitter output terminals and delivered to the antenna, antenna transmission line, or any other impedance-matched, radio frequency load.

(1) For single sideband transmitters and other transmitters employing a reduced carrier, a suppressed carrier or a controlled carrier, used at Class D stations, transmitter power is the peak envelope power.

(2) For all transmitters other than those covered by paragraph (a)(1) of this section, the transmitter power is the carrier power.

(b) The transmitter power of a station shall not exceed the following values under any condition of modulation or other circumstances.

Class of station:	Transmitter power in watts
A	50
C—27.255 MHz	25
C—26.995–27.195 MHz	4
C—72–76 MHz	0.75
D—Carrier (where applicable)	4
D—Peak envelope power (where applicable)	12

§ 95.44 External radio frequency power amplifiers prohibited.

No external radio frequency power amplifier shall be used or attached, by connection, coupling attachment or in any other way at any Class D station.

NOTE: An external radio frequency power amplifier at a Class D station will be presumed to have been used where it is in the operator's possession or on his premises and there is extrinsic evidence of any operation of such Class D station in excess of power limitations provided under this rule part unless the operator of such equipment holds a station license in another radio service under which license the use of the said amplifier at its maximum rated output power is permitted.

§ 95.45 Frequency tolerance.

(a) Except as provided in paragraphs (b) and (c) of this section, the carrier frequency of a transmitter in this service shall be maintained within the following percentage of the authorized frequency:

Class of station	Frequency tolerance	
	Fixed and base	Mobile
A	0.00025	0.0005
C		.005
D		.005

(b) Transmitters used at Class C stations operating on authorized frequencies between 26.99 and 27.26 MHz with 2.5 watts or less mean output power, which are used solely for the control of remote objects or devices by radio (other than devices used solely as a means of attracting attention), are permitted a frequency tolerance of 0.01 percent.

(c) Class A stations operated at a fixed location used to control base stations, through use of a mobile only frequency, may operate with a frequency tolerance of 0.0005 percent.

§ 95.47 Types of emission.

(a) Except as provided in paragraph (e) of this section, Class A stations in this service will normally be authorized to transmit radiotelephony only. However, the use of tone signals or signaling devices solely to actuate receiver circuits, such as tone operated squelch or selective calling circuits, the primary function of which is to establish or establish and maintain voice communications, is permitted. The use of tone signals solely to attract attention is prohibited.

(b) [Reserved]

(c) Class C stations in this service are authorized to use amplitude tone modulation or on-off unmodulated carrier only, for the control of remote objects or devices by radio or for the remote actuation of devices which are used solely as a means of attracting attention. The transmission of any form of telegraphy, telephony or record communications by a Class C station is prohibited. Telemetering, except for the transmission of simple, short duration signals indicating the presence or absence of a condition or the occurrence of an event, is also prohibited.

(d) Transmitters used at Class D stations in this service are authorized to use amplitude voice modulation, either single or double sideband. Tone signals or signalling devices may be used only to actuate receiver circuits, such as tone operated squelch or selective calling circuits, the primary function of which is to establish or maintain voice communications. The use of any signals solely to attract attention or for the control of remote objects or devices is prohibited.

(e) Other types of emission not described in paragraph (a) of this section may be authorized for Class A citizens radio stations upon a showing of need therefor. An application requesting such authorization shall fully describe the emission desired, shall indicate the bandwidth required for satisfactory communication, and shall state the purpose for which such emission is required. For information regarding the classification of emissions and the calculation of bandwidth, reference should be made to Part 2 of this chapter.

§ 95.49 Emission limitations.

(a) Each authorization issued to a Class A citizens radio station will show, as a prefix to the classification of the authorized emission, a figure specifying the maximum bandwidth to be occupied by the emission.

(b) [Reserved]

(c) The authorized bandwidth of the emission of any transmitter employing amplitude modulation shall be 8 kHz for double sideband, 4 kHz for single sideband and the authorized bandwidth of the emission of transmitters employing frequency or phase modulation (Class F2 or F3) shall be 20 kHz. The use of Class F2 and F3 emissions in the frequency band 26.96–27.28 MHz is not authorized.

(d) The mean power of emissions shall be attenuated below the mean power of the transmitter in accordance with the following schedule:

(1) When using emissions other than single sideband:

(i) On any frequency removed from the center of the authorized bandwidth by more than 50 percent up to and including 100 percent of the authorized bandwidth: at least 25 decibels;

(ii) On any frequency removed from the center of the authorized bandwidth by more than 100 percent up to and including 250 percent of the authorized bandwidth: At least 35 decibels;

(2) When using single sideband emissions:

(i) On any frequency removed from the center of the authorized bandwidth by more than 50 percent up to and including 150 percent of the authorized bandwidth: At least 25 decibels:

(ii) On any frequency removed from the center of the authorized bandwidth by more than 150 percent up to and including 250 percent of the authorized bandwidth: At least 35 decibels;

(3) On any frequency removed from the center of the authorized bandwidth by more than 250 percent of the authorized bandwidth: At least 43 plus 10 $\log_{10}$ (mean power in watts) decibels.

(e) When an unauthorized emission results in harmful interference, the Commission may, in its discretion, require appropriate technical changes in equipment to alleviate the interference.

§ 95.51 Modulation requirements.

(a) When double sideband, amplitude modulation is used for telephony, the modulation percentage shall be sufficient to provide efficient communication and shall not exceed 100 percent.

(b) Each transmitter for use in Class D stations, other than single sideband, suppressed carrier, or controlled carrier, for which type acceptance is requested after May 24, 1974, having more than 2.5 watts maximum output power shall be equipped with a device which automatically prevents modulation in excess of 100 percent on positive and negative peaks.

(c) The maximum audio frequency required for satisfactory radiotelephone intelligibility for use in this service is considered to be 3000 Hz.

(d) Transmitters for use at Class A stations shall be provided with a device which automatically will prevent greater than normal audio level from causing modulation in excess of that specified in this subpart; *Provided, however*, That the requirements of this paragraph shall not apply to transmitters authorized at mobile stations and having an output power of 2.5 watts or less.

(e) Each transmitter of a Class A station which is equipped with a modulation limiter in accordance with the provisions of paragraph (d) of this section shall also be equipped with an audio low-pass filter. This audio low-pass filter shall be installed between the modulation limiter and the modulated stage and, at audio frequencies between 3 kHz and 20 kHz, shall have an attenuation greater than the attenuation at 1 kHz by at least:

$$60 \log_{10} (f/3) \text{ decibels}$$

where "f" is the audio frequency in kHz. At audio frequencies above 20 kHz, the attentuation shall be at least 50 decibels greater than the attenuation at 1 kHz.

(f) Simultaneous amplitude modulation and frequency or phase modulation of a transmitter is not authorized.

(g) The maximum frequency deviation of frequency modulated transmitters used at Class A stations shall not exceed ±5 kHz.

§ 95.53 Compliance with technical requirements.

(a) Upon receipt of notification from the Commission of a deviation from the technical requirements of the rules in this part, the radiations of the transmitter involved shall be suspended immediately, except for necessary tests and adjustments, and shall not be resumed until such deviation has been corrected.

(b) When any citizens radio station licensee receives a notice of violation indicating that the station has been operated contrary to any of the provisions contained in Subpart C of this part, or where it otherwise appears that operation of a station in this service may not be in accordance with applicable technical standards, the Commission may require the licensee to conduct such tests as may be necessary to determine whether the equipment is capable of meeting these standards and to make such adjustments as may be necessary to assure compliance therewith. A licensee who is notified that he is required to conduct such tests and/or make adjustments must, within the time limit specified in the notice, report to the Commission the results thereof.

(c) All tests and adjustments which may be required in accordance with paragraph (b) of this section shall be made by, or under the immediate supervision of, a person holding a first- or second-class commercial operator license, either radiotelephone or radio telegraph as may be appropriate for the type of emission employed. In each case, the report which is submitted to the Commission shall be signed by the licensed commercial operator. Such report shall describe the results of the tests and adjustments, the test equipment and procedures used, and shall state the type, class, and serial number of the operator's license. A copy of this report shall also be kept with the station records.

§ 95.55 Acceptability of transmitters for licensing.

Transmitters type approved or type accepted for use under this part are included in the Commission's Radio Equipment List. Copies of this list are available for public reference at the Commission's Washington, D.C., offices and field offices. The requirements for transmitters which may be operated under a license in this service are set forth in the following paragraphs.

(a) Class A stations: All transmitters shall be type accepted.

(b) Class C stations:

(1) Transmitters operated in the band 72–76 MHz shall be type accepted.

(2) All transmitters operated in the band 26.99–27.26 MHz shall be type approved, type accepted or crystal controlled.

(c) Class D Stations:

(1) All transmitters first licensed, or marketed as specified in § 2.805 of this chapter, prior to November 22, 1974, shall be type accepted or crystal controlled.

(2) All transmitters first licensed, or marketed as specified in § 2.803 of this chapter, on or after November 22, 1974, shall be type accepted.

(3) Effective November 23, 1978, all transmitters shall be type accepted.

(4) Transmitters which are equipped to operate on any frequency not included in § 95.41(d)(1) may not be installed at, or used by, any Class D station unless there is a station license posted at the transmitter location, or a transmitter identification card (FCC Form 452–C) attached to the transmitter, which indicates that operation of the transmitter on such frequency has been authorized by the Commission.

(d) With the exception of equipment type approved for use at a Class C station, all transmitting equipment authorized in this service shall be crystal controlled.

(e) No controls, switches or other functions which can cause operation in violation of the technical regulations of this part shall be accessible from the operating panel or exterior to the cabinet enclosing a transmitter authorized in this service.

§ 95.57 Procedure for type acceptance of equipment.

(a) Any manufacturer of a transmitter built for use in this service, except noncrystal controlled transmitters for use at Class C stations, may request type acceptance for such transmitter in accordance with the type acceptance requirements of this part, following the type acceptance procedure set forth in Part 2 of this chapter.

(b) Type acceptance for an individual transmitter may also be requested by an applicant for a station authorization by following the type acceptance procedures set forth in Part 2 of this chapter. Such transmitters, if accepted, will not normally be included on the Commission's "Radio Equipment List", but will be individually enumerated on the station authorization.

(c) Additional rules with respect to type acceptance are set forth in Part 2 of this chapter. These rules include information with respect to withdrawal of type acceptance, modification of type-accepted equipment, and limitations on the findings upon which type acceptance is based.

(d) Transmitters equipped with a frequency or frequencies not listed in § 95.41(d)(1) will not be type accepted for use at Class D stations unless the transmitter is also type accepted for use in the service in which the frequency is authorized, if type acceptance in that service is required.

§ 95.58 Additional requirements for type acceptance.

(a) All transmitters shall be crystal controlled.

(b) Except for transmitters type accepted for use at Class A stations, transmitters shall not include any provisions for increasing power to levels in excess of the pertinent limits specified in Section 95.43.

(c) In addition to all other applicable technical requirements set forth in this part, transmitters for which type acceptance is requested after May 24, 1974, for use at Class D stations shall comply with the following:

(1) Single sideband transmitters and other transmitters employing reduced, suppressed or controlled carrier shall include a means for automatically preventing the transmitter power from exceeding either the maximum permissible peak envelope power or the rated peak envelope power of the transmitter, whichever is lower.

(2) Multi-frequency transmitters shall not provide more than 23 transmitting frequencies, and the frequency selector shall be limited to a single control.

(3) Other than the channel selector switch, all transmitting frequency determining circuitry, including crystals, employed in Class D station equipment shall be internal to the equipment and shall not be accessible from the exterior of the equipment cabinet or operating panel.

(4) Single sideband transmitters shall be capable of transmitting on the upper sideband. Capability for transmission also on the lower sideband is permissible.

(5) The total dissipation ratings, established by the manufacturer of the electron tubes or semiconductors which supply radio frequency power to the antenna terminals of the transmitter, shall not exceed 10 watts. For electron tubes, the rating shall be the Intermittent Commercial and Amateur Service (ICAS plate dissipation value if established. For semiconductors, the rating shall be the collector or device dissipation value, whichever is greater, which may be temperature de-rated to not more than 50°C.

(d) Only the following external transmitter controls, connections or devices will normally be permitted in transmitters for which type acceptance is requested after May 24, 1974, for use at Class D stations. Approval of additional controls, connections or devices may be given after consideration of the function to be performed by such additions.

(1) Primary power connection. (Circuitry or devices such as rectifiers, transformers, or inverters which provide the nominal rated transmitter primary supply voltage may be used without voiding the transmitter type acceptance.)

(2) Microphone connection.

(3) Radio frequency output power connection.

(4) Audio frequency power amplifier output connector and selector switch.

(5) On-off switch for primary power to transmitter. May be combined with receiver controls such as the receiver on-off switch and volume control.

(6) Upper-lower sideband selector; for single sideband transmitters only.

(7) Selector for choice of carrier level; for single sideband transmitters only. May be combined with sideband selector.

(8) Transmitting frequency selector switch.

(9) Transmit-receive switch.

(10) Meter(s) and selector switch for monitoring transmitter performance.

(11) Pilot lamp or meter to indicate the presence of radio frequency output power or that transmitter control circuits are activated to transmit.

(e) An instruction book for the user shall be furnished with each transmitter sold and one copy (a draft or preliminary copy is acceptable providing a final copy is furnished when completed) shall be forwarded to the Commission with each request for type acceptance or type approval. The book shall contain all information necessary for the proper installation and operation of the transmitter including:

(1) Instructions concerning all controls, adjustments and switches which may be operated or adjusted without causing violation of technical regulations of this part;

(2) Warnings concerning any adjustment which, according to the rules of this part, may be made only by, or under the immediate supervision of, a person holding a commercial first or second class radio operator license;

(3) Warnings concerning the replacement or substitution of crystals, tubes or other components which could cause violation of the technical regulations of this part and of the type acceptance or type approval requirements of Part 2 of this chapter.

(4) Warnings concerning licensing requirements and details concerning the application procedures for licensing.

§ 95.59 Submission of noncrystal controlled Class C station transmitters for type approval.

Type approval of noncrystal controlled transmitters for use at Class C stations in this service may be requested in accordance with the procedure specified in Part 2 of this chapter.

§ 95.61 Type approval of receiver-transmitter combinations.

Type approval will not be issued for transmitting equipment for operation under this part when such equipment is enclosed in the same cabinet, is constructed on the same chassis in whole or in part, or is identified with a common type or model number with a radio receiver, unless such receiver has been certificated to the Commission as complying with the requirements of Part 15 of this chapter.

§ 95.63 Minimum equipment specifications.

Transmitters submitted for type approval in this service shall be capable of meeting the technical specifications contained in this part, and in addition, shall comply with the following:

(a) Any basic instructions concerning the proper adjustment, use, or operation of the equipment that may be necessary shall be attached to the equipment in a suitable manner and in such positions as to be easily read by the operator.

(b) A durable nameplate shall be mounted on each transmitter showing the name of the manufacturer, the type or model designation, and providing suitable space for permanently displaying the transmitter serial number, FCC type approval number, and the class of station for which approved.

(c) The transmitter shall be designed, constructed, and adjusted by the manufacturer to operate on a frequency or frequencies available to the class of station for which type approval is sought. In designing the equipment, every reasonable precaution shall be taken to protect the user from high voltage shock and radio frequency burns. Connections to batteries (if used) shall be made in such a manner as to permit replacement by the user without causing improper operation of the transmitter. Generally accepted modern engineering principles shall be utilized in the generation of radio frequency currents so as to guard against unnecessary interference to other services. In cases of harmful interference arising from the design, construction, or operation of the equipment, the Commission may require appropriate technical changes in equipment to alleviate interference.

(d) Controls which may effect changes in the carrier frequency of the transmitter shall not be accessible from the exterior of any unit unless such accessibility is specifically approved by the Commission.

§ 95.65 Test procedure.

Type approval tests to determine whether radio equipment meets the technical specifications contained in this part will be conducted under the following conditions:

(a) Gradual ambient temperature variations from 0° to 125° F.

(b) Relative ambient humidity from 20 to 95 percent. This test will normally consist of subjecting the equipment for at least three consecutive periods of 24 hours each, to a relative ambient humidity of 20, 60, and 95 percent, respectively, at a temperature of approximately 80° F.

(c) Movement of transmitter or objects in the immediate vicinity thereof.

(d) Power supply voltage variations normally to be encountered under actual operating conditions.

(e) Additional tests as may be prescribed, if considered necessary or desirable.

§ 95.67 Certificate of type approval.

A certificate or notice of type approval, when issued to the manufacturer of equipment intended to be used or operated in the Citizens Radio Service, constitutes a recognition that on the basis of the test made, the particular type of equipment appears to have the capability of functioning in accordance with the technical specifications and regulations contained in this part: *Provided*, That all such additional equipment of the same type is properly constructed, maintained, and operated: *And provided further*, That no change whatsoever is made in the design or construction of such equipment except upon specific approval by the Commission.

SUBPART D—STATION OPERATING REQUIREMENTS

§ 95.83 Prohibited uses.

(a) A Citizens radio station shall not be used:

(1) For engaging in radio communications as a hobby or diversion, i.e., operating the radio station as an activity in and of itself.

NOTE: The following are typical, but not all inclusive, examples of the types of communications evidencing a use of Citizens radio as a hobby or diversion which are prohibited under this rule:

"You want to give me your handle and I'll ship you out a card the first thing in the morning;" or "Give me your 10-20 so I can ship you some wallpaper." (Communications to other licensees for the purpose of exchanging so-called "QSL" cards.)

"I'm just checking to see who is on the air."

"Just calling to see if you can hear me. I'm at Main and Broadway."

"Just heard your call sign and thought I'd like to get acquainted;" or "Just passing through and heard your call sign so I thought I'd give you a shout."

"Just sitting here copying the mail and thought I'd give you a call to see how you were doing." (Referring to an intent to communicate based solely on hearing another person engaged in the use of his radio.)

"My 10-20 is Main and Broad Streets. Thought I'd call so I can see how well this new rig is getting out."

"Got a new mike on this rig and thought I'd give you a call to find out how my modulation is."

"Just thought I would give you a shout and let you know I am still around. Thanks for coming back."

"Clear with Venezuela. Just thought I'd let you know I was copying you up here."

"Thought I'd give you a shout and see if you knew where the unmodulated carrier was coming from."

"Just thought I'd give you a call to find out how the skip is coming in over at your location."

"Go ahead breaker. What kind of a rig are you using? Come back with your 10-20."

(2) For any purpose, or in connection with any activity, which is contrary to Federal, State, or local law.

(3) For the transmission of communications containing obscene, indecent, or profane words, language, or meaning.

(4) To carry communications for hire, whether the remuneration or benefit received is direct or indirect.

(5) To communicate with stations authorized or operated under the provisions of other parts of this chapter, with unlicensed stations, or with U.S. Government or foreign stations (other than as provided in Subpart E of this part) except for communications pursuant to §§ 95.85(b) and 95.121 and, in the case of Class A stations, for communications with U.S. Government stations in those cases which require cooperation or coordination of activities.

(6) For any communication not directed to specific stations or persons, except for: (i) Emergency and civil defense communications as provided in §§ 95.85 (b) and 95.121, respectively, (ii) test transmissions pursuant to § 95.93, and (iii) communications from a mobile unit to other units or stations for the sole purpose of requesting routing directions, assistance to disabled vehicles or vessels, information concerning the availability of food or lodging, or any other assistance necessary to a licensee in transit.

(7) To convey program material for retransmission, live or delayed, on a broadcast facility.

NOTE: A Class A or Class D station may be used in connection with the administrative, engineering, or maintenance activities of a broadcasting station: a Class A or Class C station may be used for control functions by radio which do not involve the transmission of program material: and a Class A or Class D station may be used in the gathering of news items or preparation of programs: *Provided*, That the actual or recorded transmissions of the Citizens radio station are not broadcast at any time in whole or in part.

(8) To interfere maliciously with the communications of another station.

(9) For the direct transmission of any material to the public through public address systems or similar means.

(10) To transmit superfluous communications, i.e., any transmissions which are not necessary to communications which are permissible.

(11) For the transmission of music, whistling, sound effects, or any material for amusement or entertainment purposes, or solely to attract attention.

(12) To transmit the word "MAYDAY" or other international distress signals, except when a ship, aircraft, or other vehicle is threatened by grave and imminent danger and requests immediate assistance.

(13) For transmitting communications to stations of other licensees which relate to the technical performance, capabilities, or testing of any transmitter or other radio equipment, including transmissions concerning the signal strength or frequency stability of a transmitter, except as necessary to establish or maintain the specific communication.

(14) For relaying messages or transmitting communications for a person other than the licensee or members of his immediate family except: (i) Communications transmitted pursuant to §§ 95.85(b), 95.87(b)(7), and 95.121; (ii) upon specific prior Commission approval, communications between Citizens radio service stations at fixed locations where public telephone service is not provided; and (iii) communications reporting locally observed traffic conditions directed to persons engaged directly or indirectly in furnishing traffic condition information to the motoring public via broadcast facilities.

(15) For advertising or soliciting the sale of any goods or services.

(16) For transmitting messages in other than plain language. Abbreviations, including nationally or internationally recognized operating signals, may be used only if a list of all such abbreviations and their meaning is kept in the station records and made available to any Commission representative on demand.

(b) A Class D station may not be used to communicate with, or attempt to communicate with, any unit of the same or another station over a distance of more than 150 miles.

(c) A licensee of a Citizens radio station who is engaged in the business of selling Citizens radio transmitting equipment shall not allow a customer to operate under his station license. In addition, all communications by the licensee for the purpose of demonstrating such equipment shall consist only of brief messages addressed to other units of the same station.

§ 95.85 Emergency and assistance to motorist use.

(a) All Citizens radio stations shall give priority to the emergency communications of other stations which involve the immediate safety of life of individuals or the immediate protection of property.

(b) Any station in this service may be utilized during an emergency involving the immediate safety of life of individuals or the immediate protection of property for the transmission of emergency communications. It may also be used to transmit communications necessary to render assistance to a motorist.

(1) When used for transmission of emergency communications certain provisions in this part concerning use of frequencies (§ 95.41(d)) ; prohibited uses (§ 95.83(a) (5), (6), and (14)) ; operation by or on behalf of persons other than the licensee (§ 95.87) ; and duration of transmissions (§ 95.91 (a) and (b)) shall not apply.

(2) When used for transmission of communications necessary to render assistance to a motorist, the provisions of this part concerning directing communications to specific persons or stations (§ 95.83(a) (6)) ; transmitting messages for other persons (§ 95.83(a) (14)) ; and duration of transmissions (§ 95.91(b)) shall not apply.

(3) The exemptions granted from certain rule provisions in subparagraphs (1) and (2) of this paragraph may be rescinded by the Commission at its discretion.

(c) If the emergency use under paragraph (b) of this section extends over a period of 12 hours or more, notice shall be sent to the Commission in Washington, D.C., as soon as it is evident that the emergency has or will exceed 12 hours. The notice should include the identity of the stations participating, the nature of the emergency, and the use made of the stations. A single notice covering all participating stations may be submitted.

§ 95.87 Operation by, or on behalf of, persons other than the licensee.

(a) Transmitters authorized in this service must be under the control of the licensee at all times. A licensee shall not transfer, assign, or dispose of, in any manner, directly or indirectly, the operating authority under his station license, and shall be responsible for the proper operation of all units of the station.

(b) Citizens radio stations may be operated only by the following persons, except as provided in paragraph (c) of this section:

(1) The licensee;

(2) Members of the licensee's immediate family living in the same household;

(3) The partners, if the licensee is a partnership, provided the communications relate to the business of the partnership;

(4) The members, if the licensee is an unincorporated association, provided the communications relate to the business of the association;

(5) Employees of the licensee only while acting within the scope of their employment;

(6) Any person under the control or supervision of the licensee when the station is used solely for the control of remote objects or devices, other than devices used only as a means of attracting attention; and

(7) Other persons, upon specific prior approval of the Commission shown on or attached to the station license, under the following circumstances:

(i) Licensee is a corporation and proposes to provide private radiocommunication facilities for the transmission of messages or signals by or on behalf of its parent corporation, another subsidiary of the parent corporation, or its own subsidiary. Any remuneration or compensation received by the licensee for the use of the radiocommunication facilities shall be governed by a contract entered into by the parties concerned and the total of the compensation shall not exceed the cost of providing the facilities. Records which show the cost of service and its nonprofit or cost-sharing basis shall be maintained by the licensee.

(ii) Licensee proposes the shared or cooperative use of a Class A station with one or more other licensees in this service for the purpose of communicating on a regular basis with units of their respective Class A stations, or with units of other Class A stations if the communications transmitted are otherwise permissible. The use of these private radiocommunication facilities shall be conducted pursuant to a written contract which shall provide that contributions to capital and operating expense shall be made on a nonprofit, cost-sharing basis, the cost to be divided on an equitable basis among all parties to the agreement. Records which show the cost of service and its nonprofit, cost-sharing basis shall be maintained by the licensee. In any case, however, licensee must show a separate and independent need for the particular units proposed to be shared to fulfill his own communications requirements.

(iii) Other cases where there is a need for other persons to operate a unit of licensee's radio station. Requests for authority may be made either at the time of the filing of the application for station license or thereafter by letter. In either case, the licensee must show the nature of the proposed use and that it relates to an activity of the licensee, how he proposes to maintain control over the transmitters at all times, and why it is not appropriate for such other person to obtain a station license in his own name. The authority, if granted, may be specific with respect to the names of the persons who are permitted to operate, or may authorize operation by unnamed persons for specific purposes. This authority may be revoked by the Commission, in its discretion, at any time.

(c) An individual who was formerly a citizens radio station licensee shall not be permitted to operate any citizens radio station of the same class licensed to another person until such time as he again has been issued a valid radio station license of that class, when his license has been:

(1) Revoked by the Commission.

(2) Surrendered for cancellation after the institution of revocation proceedings by the Commission.

(3) Surrendered for cancellation after a notice of apparent liability to forfeiture has been served by the Commission.

§ 95.89 Telephone answering services.

(a) Notwithstanding the provisions of § 95.87, a licensee may install a transmitting unit of his station on the premises of a telephone answering service. The same unit may not be operated under the authorization of more than one licensee. In all cases, the licensee must enter into a written agreement with the answering service. This agreement must be kept with the licensee's station records and must provide, as a minimum, that:

(1) The licensee will have control over the operation of the radio unit at all times;

(2) The licensee will have full and unrestricted access to the transmitter to enable him to carry out his responsibilities under his license;

(3) Both parties understand that the licensee is fully responsible for the proper operation of the citizens radio station; and

(4) The unit so furnished shall be used only for the transmission of communications to other units belonging to the licensee's station.

(b) A citizens radio station licensed to a telephone answering service shall not be used to relay messages or transmit signals to its customers.

§ 95.91 Duration of transmissions.

(a) All communications or signals, regardless of their nature, shall be restricted to the minimum practicable transmission time. The radiation of energy shall be limited to transmissions modulated or keyed for actual permissible communications, tests, or control signals. Continuous or uninterrupted transmissions from a single station or between a number of communicating stations is prohibited, except for communications involving the immediate safety of life or property.

(b) Communications between or among Class D stations shall not exceed 5 consecutive minutes. At the conclusion of this 5-minute period, or upon termination of the exchange if less than 5 minutes, the station transmitting and the stations participating in the exchange shall remain silent for a period of at least 5 minutes and monitor the frequency or frequencies involved before any further transmissions are made. However, for the limited purpose of acknowledging receipt of a call, such a station or stations may answer a calling station and request that it stand by for the duration of the silent period. The time limitations contained in this paragraph may not be avoided by changing the operating frequency of the station and shall apply to all the transmissions of an operator who, under other provisions of this part, may operate a unit of more than one citizens radio station.

(c) The transmission of audible tone signals or a sequence of tone signals for the operation of the tone operated squelch or selective calling circuits in accordance with § 95.47 shall not exceed a total of 15 seconds duration. Continuous transmission of a subaudible tone for this purpose is permitted. For the purposes of this section, any tone or combination of tones having no frequency above 150 hertz shall be considered subaudible.

(d) The transmission of permissible control signals shall be limited to the minimum practicable time necessary to accomplish the desired control or actuation of remote objects or devices. The continuous radiation of energy for periods exceeding 3 minutes duration for the purpose of transmission of control signals shall be limited to control functions requiring at least one or more changes during each minute of such transmission. However, while it is actually being used to control model aircraft in flight by means of interrupted tone modulation of its carrier, a citizens radio station may transmit a continuous carrier without being simultaneously modulated if the presence or absence of the carrier also performs a control function. An exception to the limitations contained in this paragraph may be authorized upon a satisfactory showing that a continuous control signal is required to perform a control function which is necessary to insure the safety of life or property.

§ 95.93 Tests and adjustments.

All tests or adjustments of citizens radio transmitting equipment involving an external connection to the radio frequency output circuit shall be made using a nonradiating dummy antenna. However, a brief test signal, either with or without modulation, as appropriate, may be transmitted when it is necessary to adjust a transmitter to an antenna for a new station installation or for an existing installation involving a change of antenna or change of transmitters, or when necessary for the detection, measurement, and suppression of harmonic or other spurious radiation. Test transmissions using a radiating antenna shall not exceed a total of 1 minute during any 5-minute period, shall not interfere with communications already in progress on the operating frequency, and shall be properly identified as required by § 95.95, but may otherwise be unmodulated as appropriate.

§ 95.95 Station identification.

(a) The call sign of a citizens radio station shall consist of three letters followed by four digits.

(b) Each transmission of the station call sign shall be made in the English language by each unit, shall be complete, and each letter and digit shall be separately and distinctly transmitted. Only standard phonetic alphabets, nationally or internationally recognized, may be used in lieu of pronunciation of letters for voice transmission of call signs. A unit designator or special identification may be used in addition to the station call sign but not as a substitute therefor.

(c) Except as provided in paragraph (d) of this section, all transmissions from each unit of a citizens radio station shall be identified by the transmission of its assigned call sign at the beginning and end of each transmission or series of transmissions directed to or exchanged with a unit of the same station or units of other stations. Each required identification shall include not only the call sign of the station unit transmitting, but also the call sign of the station or stations with which the transmitting unit is communicating, or attempting to communicate. In the case of communications between units of the same station (intrastation), after identifying itself by its assigned call sign, the transmitting unit may identify the other units by unit designators. For communications between units of different stations (interstation), the complete sign of all stations involved must be transmitted. If the call sign of the station being called is not known, the name or trade name may be used, but when contact has been made the called station shall thereafter be identified by its call sign. Examples of proper identification procedure are set forth at the end of this paragraph. Where transmissions or exchanges of transmissions of greater length are permitted by this part, the identification shall also be transmitted at least every 15 minutes. Each transmission or exchange of transmissions conducted on different frequencies shall be fully and separately identified in accordance with the foregoing on each frequency used.

EXAMPLES OF PROPER IDENTIFICATION

Intrastation communications:

(1) Calling: "KZZ 0001 base, calling unit 2."

Response: "KZZ 0001 unit 2, to base, over."

Clearing: "KZZ 0001 base, clear with unit 2" and "KZZ 0001 unit 2, clear with base."

(2) Calling: "KZZ 0001 unit 1, calling unit 3."

Response: "KZZ 0001 unit 3, to unit 1, over."

Clearing: "KZZ 0001 unit 1, clear with unit 3" and "KZZ 0001 unit 3, clear with unit 1."

Interstation communications:

Calling: "KZZ 0001 calling KZZ 0002," or "KZZ 0001 calling KZZ 0002 unit 3" (if appropriate).

Response: "KZZ 0002 to KZZ 0001, over."

Clearing: "KZZ 0001 clear with KZZ 0002," and "KZZ 0002 clear with KZZ 0001."

(d) Unless specifically required by the station authorization, the transmissions of a citizens radio station need not be identified when the station (1) is a Class A station which automatically retransmits the information received by radio from another station which is properly identified or (2) is not being used for telephony emission.

(e) In lieu of complying with the requirements of paragraph (c) of this section, Class A base stations, fixed stations, and mobile units when communicating with base stations may identify as follows:

(1) Base stations and fixed stations of a Class A radio system shall transmit their call signs at the end of each transmission or exchange of transmissions, or once each 15-minute period of a continuous exchange of communications.

(2) A mobile unit of a Class A station communicating with a base station of a Class A radio system on the same frequency shall transmit once during each exchange of transmissions any unit identifier which is on file in the station records of such base station.

(3) A mobile unit of Class A stations communicating with a base station of a Class A radio system on a different frequency shall transmit its call sign at the end of each transmission or exchange of transmissions, or once each 15-minute period of a continuous exchange of communications.

§ 95.97 Operator license requirements.

(a) No operator license is required for the operation of a citizens radio station except that stations manually transmitting Morse Code shall be operated by the holders of a third or higher class radiotelegraph operator license.

(b) Except as provided in paragraph (c) of this section, all transmitter adjustments or tests while radiating energy during or coincident with the construction, installation, servicing, or maintenance of a radio station in this service, which may affect the proper operation of such stations, shall be made by or under the immediate supervision and responsibility of a person holding a first- or second-class commercial radio operator license, either radiotelephone or radio telegraph, as may be appropriate for the type of emission employed, and such person shall be responsible for the proper functioning of the station equipment at the conclusion of such adjustments or tests. Further, in any case where a transmitter adjustment which may affect the proper operation of the transmitter has been made while not radiating energy by a person not the holder of the required commercial radio operator license or not under the supervision of such licensed operator, other than the factory assembling or repair of equipment, the transmitter shall be checked for compliance with the technical requirements of the rules by a commercial radio operator of the proper grade before it is placed on the air.

(c) Except as provided in § 95.53 and in paragraph (d) of this section, no commercial radio operator license is required to be held by the person performing transmitter adjustments or tests during or coincident with the construction, installation, servicing, or maintenance of Class C transmitters, or Class D transmitters used at stations authorized prior to May 24, 1974: *Provided*, That there is compliance with all of the following conditions:

(1) The transmitting equipment shall be crystal-controlled with a crystal capable of maintaining the station frequency within the prescribed tolerance;

(2) The transmitting equipment either shall have been factory assembled or shall have been provided in kit form by a manufacturer who provided all components together with full and detailed instructions for their assembly by nonfactory personnel;

(3) The frequency determining elements of the transmitter, including the crystal(s) and all other components of the crystal oscillator circuit, shall have been preassembled by the manufacturer, pretuned to a specific available frequency, and sealed by the manufacturer so that replacement of any component or any adjustment which might cause off-frequency operation cannot be made without breaking such seal and thereby voiding the certification of the manufacturer required by this paragraph;

(4) The transmitting equipment shall have been so designed that none of the transmitter adjustments or tests normally performed during or coincident with the installation, servicing, or maintenance of the station, or during the normal rendition of the service of the station, or during the final assembly of kits or partially preassembled units, may reasonably be expected to result in off-frequency operation, excessive input power, overmodulation, or excessive harmonics or other spurious emissions; and

(5) The manufacturer of the transmitting equipment or of the kit from which the transmitting equipment is assembled shall have certified in writing to the purchaser of the equipment (and to the Commission upon request) that the equipment has been designed, manufactured, and furnished in accordance with the specifications contained in the foregoing subparagraphs of this paragraph. The manufacturer's certification concerning design and construction features of Class C or Class D station transmitting equipment, as required if the provisions of this paragraph are invoked, may be specific as to a particular unit of transmitting equipment or general as to a group or model of such equipment, and may be in any form adequate to assure the purchaser of the equipment or the Commission that the conditions described in this paragraph have been fulfilled.

(d) Any tests and adjustments necessary to correct any deviation of a transmitter of any Class of station in this service from the technical requirements of the rules in this part shall be made by, or under the immediate supervision of, a person holding a first- or second-class commercial operator license, either radiotelephone or radiotelegraph, as may be appropriate for the type of emission employed.

§ 95.101 Posting station license and transmitter identification cards or plates.

(a) The current authorization, or a clearly legible photocopy thereof, for each station (including units of a Class C or Class D station) operated at a fixed location shall be posted at a conspicuous place at the principal fixed location from which such station is controlled, and a photocopy of such authorization shall also be posted at all other fixed locations from which the station is controlled. If a photocopy of the authorization is posted at the principal control point, the location of the original shall be stated on that photocopy. In addition, an executed Transmitter Identification Card (FCC Form 452-C) or a plate of metal or other durable substance, legibly indicating the call sign and the licensee's name and address, shall be affixed, readily visible for inspection, to each transmitter operated at a fixed location when such transmitter is not in view of, or is not readily accessible to, the operator of at least one of the locations at which the station authorization or a photocopy thereof is required to be posted.

(b) The current authorization for each station operated as a mobile station shall be retained as a permanent part of the station records, but need not be posted. In addition, an executed Transmitter Identification Card (FCC Form 452-C) or a plate of metal or other durable substance, legibly indicating the call sign and the licensee's name and address, shall be affixed, readily visible for inspection, to each of such transmitters: *Provided*, That, if the transmitter is not in view of the location from which it is controlled, or is not readily accessible for inspection, then such card or plate shall be affixed to the control equipment at the transmitter operating position or posted adjacent thereto.

§ 95.103 Inspection of stations and station records.

All stations and records of stations in the Citizens Radio Service shall be made available for inspection upon the request of an authorized representative of the Commission made to the licensee or to his representative (see § 1.6 of this chapter). Unless otherwise stated in this part, all required station records shall be maintained for a period of at least 1 year.

§ 95.105 Current copy of rules required.

Each licensee in this service shall maintain as a part of his station records a current copy of Part 95, Citizens Radio Service, of this chapter.

§ 95.107 Inspection and maintenance of tower marking and lighting, and associated control equipment.

The licensee of any radio station which has an antenna structure required to be painted and illuminated pursuant to the provisions of section 303(q) of the Communications Act of 1934, as amended, and Part 17 of this chapter, shall perform the inspection and maintain the tower marking and lighting, and associated control equipment, in accordance with the requirements set forth in Part 17 of this chapter.

§ 95.111 Recording of tower light inspections.

When a station in this service has an antenna structure which is required to be illuminated, appropriate entries shall be made in the station records in conformity with the requirements set forth in Part 17 of this chapter.

§ 95.113 Answers to notices of violations.

(a) Any licensee who appears to have violated any provision of the Communications Act or any provision of this chapter shall be served with a written notice calling the facts to his attention and requesting a statement concerning the matter. FCC Form 793 may be used for this purpose.

(b) Within 10 days from receipt of notice or such other period as may be specified, the licensee shall send a written answer, in duplicate, direct to the office of the Commission originating the notice. If an answer cannot be sent nor an acknowledgment made within such period by reason of illness or other unavoidable circumstances, acknowledgment and answer shall be made at the earliest practicable date with a satisfactory explanation of the delay.

(c) The answer to each notice shall be complete in itself and shall not be abbreviated by reference to other communications or answers to other notices. In every instance the answer shall contain a statement of the action taken to correct the condition or omission complained of and to preclude its recurrence. If the notice relates to violations that may be due to the physical or electrical characteristics of transmitting apparatus, the licensee must comply with the provisions of § 95.53, and the answer to the notice shall state fully what steps, if any, have been taken to prevent future violations, and, if any new apparatus is to be installed, the date such apparatus was ordered, the name of the manufacturer, and the promised date of delivery. If the installation of such apparatus requires a construction permit, the file number of the application shall be given, or if a file number has not been assigned by the Commission, such identification shall be given as will permit ready identification of the application. If the notice of violation relates to lack of attention to or improper operation of the transmitter, the name and license number of the operator in charge, if any, shall also be given.

§ 95.115 False signals.

No person shall transmit false or deceptive communications by radio or identify the station he is operating by means of a call sign which has not been assigned to that station.

§ 95.117 Station location.

(a) The specific location of each Class A base station and each Class A fixed station and the specific area of operation of each Class A mobile station shall be indicated in the application for license. An authorization may be granted for the operation of a Class A base station or fixed station in this service at unspecified temporary fixed locations within a specified general area of operation. However, when any unit or units of a base station or fixed station authorized to be operated at temporary locations actually remains or is intended to remain at the same location for a period of over a year, application for separate authorization specifying the fixed location shall be made as soon as possible but not later than 30 days after the expiration of the 1-year period.

(b) A Class A mobile station authorized in this service may be used or operated anywhere in the United States subject to the provisions of paragraph (d) of this section: *Provided*, That when the area of operation is changed for a period exceeding 7 days, the following procedure shall be observed:

(1) When the change of area of operation occurs inside the same Radio District, the Engineer in Charge of the Radio District involved and the Commission's office, Washington, D.C., 20554, shall be notified.

(2) When the station is moved from one Radio District to another, the Engineers in Charge of the two Radio Districts involved and the Commission's office, Washington, D.C., 20554, shall be notified.

(c) A Class C or Class D mobile station may be used or operated anywhere in the United States subject to the provisions of paragraph (d) of this section.

(d) A mobile station authorized in this service may be used or operated on any vessel, aircraft, or vehicle of the United States: *Provided*, That when such vessel, aircraft, or vehicle is outside the territorial limits of the United States, the station, its operation, and its operator shall be subject to the governing provisions of any treaty concerning telecommunications to which the United States is a party, and when within the territorial limits of any foreign country, the station shall be subject also to such laws and regulations of that country as may be applicable.

§ 95.119 Control points, dispatch points, and remote control.

(a) A control point is an operating position which is under the control and supervision of the licensee, at which a person immediately responsible for the proper operation of the transmitter is stationed, and at which adequate means are available to aurally monitor all transmissions and to render the transmitter inoperative. Each Class A base or fixed station shall be provided with a control point, the location of which will be specified in the license. The location of the control point must be the same as the transmitting equipment unless the application includes a request for a different location. Exception to the requirement for a control point may be made by the Commission upon specific request and justification therefor in the case of certain unattended Class A stations employing special emissions pursuant to § 95.47(e). Authority for such exception must be shown on the license.

(b) A dispatch point is any position from which messages may be transmitted under the supervision of the person at a control point who is responsible for the proper operation of the transmitter. No authorization is required to install dispatch points.

(c) Remote control of a Citizens radio station means the control of the transmitting equipment of that station from any place other than the location of the transmitting equipment, except that direct mechanical control or direct electrical control by wired connections of transmitting equipment from some other point on the same premises, craft, or vehicle shall not be considered remote control. A Class A base or fixed station may be authorized to be used or operated by remote control from another fixed location or from mobile units: *Provided*, That adequate means are available to enable the person using or operating the station to render the transmitting equipment inoperative from each remote control position should improper operation occur.

(d) Operation of any Class C or Class D station by remote control is prohibited.

§ 95.121 Civil defense communications.

A licensee of a station authorized under this part may use the licensed radio facilities for the transmission of messages relating to civil defense activities in connection with official tests or drills conducted by, or actual emergencies proclaimed by, the civil defense agency having jurisdiction over the area in which the station is located: *Provided*, That:

(a) The operation of the radio station shall be on a voluntary basis.

(b) [Reserved]

(c) Such communications are conducted under the direction of civil defense authorities.

(d) As soon as possible after the beginning of such use, the licensee shall send notice to the Commission in Washington, D.C., and to the Engineer in Charge of the Radio District in which the station is located, stating the nature of the communications being transmitted and the duration of the special use of the station. In addition, the Engineer in Charge shall be notified as soon as possible of any change in the nature of or termination of such use.

(e) In the event such use is to be a series of pre-planned tests or drills of the same or similar nature which are scheduled in advance for specific times or at certain intervals of time, the licensee may send a single notice to the Commission in Washington, D.C., and to the Engineer in Charge of the Radio District in which the station is located, stating the nature of the communications to be transmitted, the duration of each such test, and the times scheduled for such use. Notice shall likewise be given in the event of any change in the nature of or termination of any such series of tests.

(f) The Commission may, at any time, order the discontinuance of such special use of the authorized facilities.

SUBPART E—OPERATION OF CITIZENS RADIO STATIONS IN THE UNITED STATES BY CANADIANS

§ 95.131 Basis, purpose and scope.

(a) The rules in this subpart are based on, and are applicable solely to the agreement (TIAS #6931) between the United States and Canada, effective July 24, 1970, which permits Canadian stations in the General Radio Service to be operated in the United States.

(b) The purpose of this subpart is to implement the agreement (TIAS #6931) between the United States and Canada by prescribing rules under which a Canadian licensee in the General Radio Service may operate his station in the United States.

§ 95.133 Permit required.

Each Canadian licensee in the General Radio Service desiring to operate his radio station in the United States, under the provisions of the agreement (TIAS #6931), must obtain a permit for such operation from the Federal Communications Commission. A permit for such operation shall be issued only to a person holding a valid license in the General Radio Service issued by the appropriate Canadian governmental authority.

§ 95.135 Application for permit.

(a) Application for a permit shall be made on FCC Form 410–B. Form 410–B may be obtained from the Commission's Washington, D.C., office or from any of the Commission's field offices. A separate application form shall be filed for each station or transmitter desired to be operated in the United States.

(b) The application form shall be completed in full in English and signed by the applicant. The application must be filed by mail or in person with the Federal Communications Commission, Gettysburg, Pa. 17325, U.S.A. To allow sufficient time for processing, the application should be filed at least 60 days before the date on which the applicant desires to commence operation.

(c) The Commission, at its discretion, may require the Canadian licensee to give evidence of his knowledge of the Commission's applicable rules and regulations. Also the Commission may require the applicant to furnish any additional information it deems necessary.

§ 95.137 Issuance of permit.

(a) The Commission may issue a permit under such conditions, restrictions and terms as it deems appropriate.

(b) Normally, a permit will be issued to expire 1 year after issuance but in no event after the expiration of the license issued to the Canadian licensee by his government.

(c) If a change in any of the terms of a permit is desired, an application for modification of the permit is required. If operation beyond the expiration date of a permit is desired an application for renewal of the permit is required. Application for modification or for renewal of a permit shall be filed on FCC Form 410–B.

(d) The Commission, in its discretion, may deny any application for a permit under this subpart. If an application is denied, the applicant will be notified by letter. The applicant may, within 30 days of the mailing of such letter, request the Commission to reconsider its action.

§ 95.139 Modification or cancellation of permit.

At any time the Commission may, in its discretion, modify or cancel any permit issued under this subpart. In this event, the permittee will be notified of the Commission's action by letter mailed to his mailing address in the United States and the permittee shall comply immediately. A permittee may, within 30 days of the mailing of such letter, request the Commission to reconsider its action. The filing of a request for reconsideration shall not stay the effectiveness of that action, but the Commission may stay its action on its own motion.

§ 95.141 Possession of permit.

The current permit issued by the Commission, or a photocopy thereof, must be in the possession of the operator or attached to the transmitter. The license issued to the Canadian licensee by his government must also be in his possession while he is in the United States.

§ 95.143 Knowledge of rules required.

Each Canadian permittee, operating under this subpart, shall have read and understood this Part 95, Citizens Radio Service.

§ 95.145 Operating conditions.

(a) The Canadian licensee may not under any circumstances begin operation until he has received a permit issued by the Commission.

(b) Operation of station by a Canadian licensee under a permit issued by the Commission must comply with all of the following:

(1) The provisions of this subpart and of Subparts A through D of this part.

(2) Any further conditions specified on the permit issued by the Commission.

§ 95.147 Station identification.

The Canadian licensee authorized to operate his radio station in the United States under the provisions of this subpart shall identify his station by the call sign issued by the appropriate authority of the government of Canada followed by the station's geographical location in the United States as nearly as possible by city and state.

Changes effective February 5, 1975.

22. Section 95.7 is revised to read as follows:

§ 95.7 General citizenship requirements.

A station license shall not be granted to or held by a foreign government or a representative thereof.

23. Section 95.14 is added to read as follows:

§ 95.14 Mailing address furnished by licensee.

Each application shall set forth and each licensee shall furnish the Commission with an address in the United States to be used by the Commission in serving documents or directing correspondence to that licensee. Unless any licensee advises the Commission to the contrary, the address contained in the licensee's most recent application will be used by the Commission for this purpose.

Changes effective September 15, 1975.

1. Section 95.37(c) and (c)(3) are revised to read as follows:

§ 95.37 Limitations on antenna structures.

*   *   *   *   *

(c) All antennas (both receiving and transmitting) and supporting structures associated or used in conjunction with a Class C or D Citizens Radio Station operated from a fixed location must comply with at least one of the following:

(1) * * *

(2) * * *

(3) The antenna is mounted on the transmitting antenna structure of another authorized radio station and exceeds neither 60 feet above ground level nor the height of the antenna supporting structure of the other station; or

*   *   *   *   *

2. In § 95.41(d) (1) is revised, paragraph (2) is deleted, paragraph (3) is redesignated as paragraph (2), and a new paragraph (3) is added to read as follows:

§ 95.41 Frequencies available.

*   *   *   *   *

(d) * * *

(1) The following frequencies, commonly known as channels, may be use for

communication between units of the same station (intrastation) or different stations (interstation):

MHz	Channel
26.965	1
26.975	2
26.985	3
27.005	4
27.015	5
27.025	6
27.035	7
27.055	8
27.075	10
27.105	12
27.115	13
27.125	14
27.135	15
27.155	16
27.165	17
27.175	18
27.185	19
27.205	20
27.215	21
27.225	22
27.255	23

(2) * * * (text of current subparagraph (3))

(3) The frequency 27.085 MHz (Channel 11) shall be used only as a calling frequency for the sole purpose of establishing communications and moving to another frequency (channel) to conduct communications.

* * * * *

3. A new Section 95.81 is added to read as follows:

§ 95.81 Permissible communications.

Stations licensed in the Citizens Radio Service are authorized to transmit the following types of communications:

(a) Communications to facilitate the personal or business activities of the licensee.

(b) Communication relating to:

(1) The immediate safety of life or the immediate protection of property in accordance with § 95.85.

(2) The rendering of assistance to a motorist, mariner or other traveler.

(3) Civil defense activities in accordance with § 95.121.

(4) Other activities only as specifically authorized pursuant to § 95.87.

(c) Communications with stations authorized in other radio services except as prohibited in § 95.83(a)(3).

4. Section 95.83(a) and headnote are amended to read as follows:

§ 95.83 Prohibited communications.

(a) A citizens radio station shall not be used:

(1) For any purpose, or in connection with any activity, which is contrary to Federal, State, or local law.

(2) For the transmission of communications containing obscene, indecent, profane words, language, or meaning.

(3) To communicate with an Amateur Radio Service station, an unlicensed station, or foreign stations (other than as provided in Subpart E of this part) except for communications pursuant to §§ 95.85(b) and 95.121.

(4) To convey program material for retransmission, live or delayed, on a broadcast facility. Note: A Class A or Class D station may be used in connection with administrative, engineering, or maintenance activities of a broadcasting station; a Class A or Class C station may be used for control functions by radio which do not involve the transmission of program material; and a Class A or Class D station may be used in the gathering of news items or preparation of programs: Provided, that the actual or recorded transmissions of the Citizens radio station are not broadcast at any time in whole or in part.

(5) To intentionally interfere with the communications of another station.

(6) For the direct transmission of any material to the public through a public address system or similar means.

(7) For the transmission of music, whistling, sound effects, or any material for amusement or entertainment purposes, or solely to attract attention.

(8) To transmit the word "MAYDAY" or other international distress signals, except when the station is located in a ship, aircraft, or other vehicle which is threatened by grave and imminent danger and requests immediate assistance.

(9) For advertising or soliciting the sale of any goods or services.

(10) For transmitting messages in other than plain language. Abbreviations including nationally or internationally recognized operating signals, may be

used only if a list of all such abbreviations and their meaning is kept in the station records and made available to any Commission representative on demand.

(11) To carry on communications for hire, whether the remuneration or benefit received is direct or indirect.

* * * * *

5. Section 95.85(b) (1) and (2) are revised to read as follows:

§ 95.85 Emergency and assistance to motorist use.

* * * * *

(1) When used for transmission of emergency communications certain provisions in this part concerning use of frequencies (§ 95.41(d)); prohibited uses (§ 95.83(a)(3)); operation by or on behalf of persons other than the licensee (§ 95.87); and duration of transmissions (§ 95.91(a) and (b)) shall not apply.

(2) When used for transmissions of communications necessary to render assistance to a traveler, the provisions of this Part concerning duration of transmission (§ 95.91(b)) shall not apply.

* * * * *

6. Section 95.91 is amended to add new paragraphs (b) and (c) and to redesignate the present paragraphs (c) and (d), as (e) and (f), respectively.

§ 95.91 Duration of transmissions.

* * * * *

(b) All communications between Class D stations (interstation) shall be restricted to not longer than five (5) continuous minutes. At the conclusion of this 5 minute period, or the exchange of less than 5 minutes, the participating stations shall remain silent for at least one minute.

(c) All communication between units of the same Class D station (intrastation) shall be restricted to the minimum practicable transmission.

* * * * *

7. Section 95.95(c) is revised to read as follows:

§ 95.95 Station identification.

* * * * *

(c) Except as provided in paragraph (d) of this section, all transmission from each unit of a citizens radio station shall be identified by the transmission of its assigned call sign at the beginning and end of each transmission or series of transmissions, but at least at intervals not to exceed ten (10) minutes.

* * * * *

8. Section 95.119(d) is revised to read as follows:

§ 95.119 Control points, dispatch points, and remote control.

* * * * *

(d) Operation of any Class C or Class D station by remote control is prohibited except remote control by wire upon specific authorization by the Commission when satisfactory need is shown.

APPENDIX II

FCC Field Offices and Districts

FCC FIELD ENGINEERING OFFICES

Address all communications to the Engineer-in-Charge, FCC. Street addresses can be found in local telephone directories under "United States Government."

Alabama, Mobile 36602
Alaska, Anchorage (Box 644) 99501
California, Los Angeles 90012
California, San Diego 92101
California, San Francisco 94111
California, San Pedro 90731
Colorado, Denver 80202
D.C., Washington 20554
Florida, Miami 33130
Florida, Tampa 33602
Georgia, Atlanta 30303
Georgia, Savannah (Box 8004) 31402
Hawaii, Honolulu 96808
Illinois, Chicago 60604
Louisiana, New Orleans 70130
Maryland, Baltimore 21202
Massachusetts, Boston 02109
Michigan, Detroit 48226
Minnesota, St. Paul 55102
Missouri, Kansas City 64106
New York, Buffalo, 14203
New York, New York 10014
Oregon, Portland 97204
Pennsylvania, Philadelphia 19106
P. Rico, San Juan (Box 2987) 00903
Texas, Beaumont 77701
Texas, Dallas 75202
Texas, Houston 77002
Virginia, Norfolk 23510
Washington, Seattle 98104

FCC RADIO DISTRICTS

The United States is divided into 24 radio districts for purposes of license administration. The divisions are not necessarily by state lines; hence, a portion of a state may be in one district and the remainder in another. By referring to the following table, the district for any location can be determined.

District	States	Territory Within District Counties
1	Connecticut	All counties.
	Maine	All counties.
	Massachusetts	All counties.
	New Hampshire	All counties.
	Rhode Island	All counties.
	Vermont	All counties.
2	New Jersey	Bergen, Essex, Hudson, Hunterdon, Mercer, Middlesex, Monmouth, Morris, Passaic, Somerset, Sussex, Union, and Warren.
	New York	Albany, Bronx, Columbia, Delaware, Dutchess, Greene, Kings, Nassau, New York, Orange, Putnam, Queens, Rensselaer, Richmond, Rockland, Schenectady, Suffolk, Sullivan, Ulster, and Westchester.
3	Delaware	New Castle.
	New Jersey	Atlantic, Burlington, Camden, Cape May, Cumberland, Gloucester, Ocean, and Salem.
	Pennsylvania	Adams, Berks, Bucks, Carbon, Chester, Cumberland, Dauphin, Delaware, Lancaster, Lebanon, Lehigh, Monroe, Montgomery, Northampton, Perry, Philadelphia, Schuylkill, and York.
4	Delaware	Kent and Sussex.
	Maryland	All except District 24.
	Virginia	Clarke, Fairfax all except District 24, Fauquier, Frederick, Loudoun, Page, Prince, William, Rappahannock, Shenandoah, and Warren.
	West Virginia	Barbour, Berkeley, Grant, Hampshire, Hardy, Harrison, Jefferson, Lewis, Marion, Mineral, Monongalia, Morgan, Pendleton, Preston, Randolph, Taylor, Tucker, Upshur.
5	North Carolina	All except District 6.
	Virginia	All except Districts 4 and 24.
6	Alabama	All except District 8.
	Georgia	All counties.
	North Carolina	Ashe, Avery, Buncombe, Burke, Caldwell, Cherokee, Clay, Cleveland, Graham, Haywood, Henderson, Jackson, McDowell, Macon, Madison, Mitchell, Polk, Rutherford, Swain, Transylvania, Watauga, Yancey.
	South Carolina	All counties.

(*Continued on next page*)

District	States	Territory Within District Counties
	Tennessee	All counties.
7	Florida	All except District 8.
8	Alabama	Baldwin and Mobile.
	Arkansas	All counties.
	Florida	Escambia.
	Louisiana	All counties.
	Mississippi	All counties.
	Texas	City of Texarkana only.
9	Texas	Angelina, Aransas, Atascosa, Austin, Bandera, Bastrop, Bee, Bexar, Blanco, Brazoria, Brazos, Brooks, Burleson, Caldwell, Calhoun, Cameron, Chambers, Colorado, Comal, DeWitt, Dimmit, Duval, Edwards, Fayette, Fort Bend, Frio, Galveston, Gillespie, Goliad, Gonzales, Grimes, Guadalupe, Hardin, Harris, Hayes, Hidalgo, Jackson, Jasper, Jefferson, Jim Hogg, Jim Wells, Karnes, Kendall, Kenedy, Kerr, Kinney, Kleberg, LaSalle, Lavaca, Lee, Liberty, Live Oak, Madison, Matagorda, Maverick, McMullen, Medina, Montgomery, Nacagdoches, Newton, Nueces, Orange, Polk, Real, Refugio, Sabine, San Augustine, San Jacinto, San Patricio, Starr, Travis, Trinity, Tyler, Uvalde, Val Verde, Victoria, Walker, Waller, Washington, Webb, Wharton, Willacy, Williamson, Wilson, Zapata, and Zavala.
10	Oklahoma	All counties.
	Texas	All except District 9 and the city of Texarkana.
11	Arizona	All counties.
	California	Imperial, Inyo, Kern, Los Angeles, Orange, Riverside, San Bernardino, San Diego, San Luis, Obispo, Santa Barbara, and Ventura.
	Nevada	Clark.
12	California	All except District 11.
	Nevada	All except Clark.
13	Idaho	All except District 14.
	Oregon	All counties.
	Washington	Clark, Cowlitz, Klickitat, Skamania, and Wahkiakum.

District	States	Territory Within District Counties
14	Idaho	Benewah, Bonner, Boundary, Clearwater, Idaho Kootenai, Latah, Lewis, Nez Perce, Shoshone.
	Montana	All counties.
	Washington	All except District 13.
15	Colorado	All counties.
	Utah	All counties.
	Wyoming	All counties.
	Nebraska	Baner, Box Butte, Cheyenne, Dawes, Deuel, Garden, Kimball, Morrill, Scotts Bluff, Sheridan, Sioux.
	New Mexico	All counties.
	South Dakota	Butte, Custer, Fall River, Lawrence, Meade, Pennington, Shannon, Washabaugh.
16	Minnesota	All counties.
	Michigan	Alger, Baraga, Chippewa, Delta, Dickinson, Gogebic, Houghton, Iron, Keweenaw, Luce, Mackinac, Marquette, Menominee, Ontonagon, and Schoolcraft.
	South Dakota	All counties except District 15.
	North Dakota	All counties.
	Wisconsin	All counties except District 18.
17	Iowa	All except District 18.
	Kansas	All counties.
	Missouri	All counties.
	Nebraska	All except District 15.
18	Illinois	All counties.
	Indiana	All counties.
	Iowa	Allamakee, Buchanan, Cedar, Clayton, Clinton, Delaware, Des Moines, Dubuque, Fayette, Henry, Jackson, Johnson, Jones, Lee, Linn, Louisa, Muscatine, Scott, Washington, and Winneshiek.
	Wisconsin	Brown, Calumet, Columbia, Crawford, Dane, Dodge, Door, Fond du Lac, Grant, Green, Iowa, Jefferson, Kenosha, Kewaunee, Lafayette, Manitowoc, Marinetee, Milwaukee, Oconto, Outagamie, Ozaukee, Racine, Richland, Rock, Sauk, Sheboygan, Walworth, Washington, Waukesha, and Winnebago.
	Kentucky	All counties except District 19.
19	Kentucky	Bath, Bell, Boone, Bourbon, Boyd, Bracken, Breathitt, Campbell, Carter,

(*Continued on next page*)

District	States	Counties Territory Within District
19–(Cont'd)		Clark, Clay, Elliott, Estill, Fayette, Fleming, Floyd, Franklin, Gallatin, Garrard, Grant, Greenup, Harlan, Harrison, Jackson, Jessamine, Johnson, Kenton, Knott, Knox, Laurel, Lawrence, Lee, Leslie, Letcher, Lewis, Lincoln, Madison, Magoffin, Martin, Mason, McCreary, Menifee, Montgomery, Morgan, Nicholas, Owen, Owsley, Pendleton, Perry, Pike, Powell, Pulaski, Robertson, Rockcastle, Rowan, Scott, Wayne, Whitley, Wolf, and Woodford.
	Ohio	All counties.
	Michigan	All counties except District 16.
	West Virginia	All counties except District 4.
20	New York	All except District 2.
	Pennsylvania	All except District 3.
21	Hawaii and outlying Pacific possessions.	
22	Puerto Rico	
	Virgin Islands	
23	Alaska	
24	District of Columbia and 10 miles beyond the boundary of the District of Columbia in each direction.	

APPENDIX III

Glossary

"A" lead. The "hot," or supply, lead on d-c radio equipment.
A-c generator. A mechanical or electrical device that produces an a-c voltage.
Adjacent-channel interference. Interference caused by equipment operating on one or both adjoining channels.
Alignment. The process of adjusting circuits to respond to desired frequencies.
Alternating current (ac). A current that reverses its direction at regular intervals.
Alternator. A device that produces an alternating (ac) voltage.
Amateur. A person licensed to operate, install, build, and experiment with transmitters, receivers, and other types of electronic equipment as a hobby. Often referred to as a "ham."
Amateur bands. Bands of frequencies allocated specifically for radio amateurs.
Ambient temperature. The temperature of the air in the surrounding area.
Ammeter. A meter designed to measure electric current.
Ampere. The unit of electric current flow.
Amplification. The process of increasing voltage and/or current.
Amplifier. A device that produces greater voltage and/or current at its output than is applied to the input.
Amplitude modulation. The process of varying the amplitude of an r-f carrier in accordance with the intelligence signal.
Antenna gain. The effectiveness of an antenna in a certain direction, compared with some standard antenna.
Atmospheric interference. Interference caused by electrical disturbances in the atmosphere.
Audio frequency. The frequency of a signal corresponding to sounds that can be heard by the human ear. The audio frequency extends from 20 to 20,000 hertz.
Automatic volume control (avc). A system in which the gain of one or more receiver stages automatically increases to compensate for reductions in signal strength.
Autotransformer. A transformer having a single winding, a tapped portion acting as the primary.
B+. The positive terminal of a d-c voltage source (battery, power supply, etc.).
B supply. The plate-voltage source for a vacuum tube.
Background noise. Random circuit or atmospheric noise heard in addition to the desired signal.
Bandwidth. The difference, in hertz between the frequency limits of a band.
Beam antenna. An antenna with a radiation or reception pattern confined to a somewhat narrow beam.
Beat frequency. A third (difference) frequency produced when two different frequencies are combined.
Bias. A counteracting voltage used to control or limit the gain of an amplifier.
Bonding. Bringing two metal objects to the same electrical potential (usually ground) by connecting them together with a heavy metal conductor.

Capacitive coupling. A method of coupling two circuits electrostatically in order to provide an a-c path for desired signals while blocking dc.
Capacitor. Two conductors separated by an insulator (dielectric).
Carrier frequency. The number of hertz of the unmodulated r-f wave produced by a transmitter.
Carrier wave. The unmodulated r-f signal produced by a transmitter.
Cathode. The electron-emitting element of a tube or other radio component.
Channel. A band within which the assigned transmitter carrier frequency and its modulation must be confined during operation.
Channel selector. A switch or dial used to select a desired channel.
Class-A amplifier. An amplifier biased so that plate current flows at all times.
Class-C amplifier. An amplifier biased so that plate current does not flow at zero input signal, and flows only during a portion of the cycle when a signal is present at the grid.
Coaxial antenna. An antenna that is a quarter-wavelength extension of the center conductor of a coaxial cable. The outer conductor is connected to and covered with a quarter-wavelength sheath, or "skirt."
Coaxial cable. A high-frequency transmission line consisting of an inner conductor and an outer shielding conductor. The two are separated by a dielectric material, and the entire assembly is covered with an insulating sheath.
Co-channel interference. Interference from an undesired signal having the same carrier frequency as the desired signal.
Collector. An electron-capturing electrode.
Continuity. The property of having a continuous d-c electrical path.
Converter. A device for changing incoming radio signals to a frequency that can be readily handled by a receiver. Also a mechanical device for changing electrical energy from one form to another—such as ac to dc.
Cps. Abbreviation for *cycles per second* (now called hertz).
Crystal. A piezoelectric material, such as natural quartz, used as the frequency-determining element in some oscillator circuits.
Cycle. One complete round of a regularly recurring event. For example, an a-c cycle consists of a movement from zero to maximum in one direction, beyond zero to maximum in the opposite direction, and then back to zero.
D-c resistance. The resistance (in ohms) a circuit or component offers to the flow of direct current.
Decibel (db). The amount the pressure of a sine-wave sound must be changed in order for such change to be barely noticeable by the average human ear.
Delayed avc. An automatic volume-control circuit that permits maximum gain on weak signals by producing the avc voltage for only those signals above a predetermined strength.
Detector. In a receiver, the stage where demodulation (detection) occurs.
Detune. To change the inductance and/or capacitance of a tuned circuit, causing it to be resonant at other than the desired frequency.
Dielectric. An insulating material between two conductors.
Diode. A device having two electrodes—a cathode for emitting electrons, and an anode for collecting them.
Dipole antenna. A two-element antenna the length of which is usually no more than half the wavelength to which it is resonant.
Direct current (dc). Current that travels in one direction only.
Directional antenna. An antenna that radiates and receives signals better in some directions than in others.
Dress. The exact placement of leads and components to minimize or eliminate undesirable feedback and other troubles.
Dummy antenna. A device used during transmitter tests and adjustments. It duplicates the electrical characteristics of the antenna, but greatly reduces or eliminates the radiation of radio waves.
Efficiency. The ratio of energy output to energy input, usually expressed as a percentage.
Electromagnetic induction. The action by which a voltage is induced in a conductor when it is exposed to an electromagnetic wave.
Electromagnetic wave. A wave consisting of electric and magnetic lines of force at right angles, such as the wave radiated from the transmitter antenna.

Emission. The radiation of a signal from the antenna into free space. Also, the giving off of electrons by an element.

Emission types. A classification of the different types of radio transmission, modulation, and other supplementary characteristics, designated by symbols such as A1, A3, etc.

Emitter. An electron-emitting element.

Federal Communications Commission (FCC). The federal agency established to regulate the operation of all communications systems.

Feedback. The return of a small amount of energy from the output of a circuit to a previous point in the circuit. In-phase feedback causes regeneration; out-of-phase, degeneration.

Filter capacitor. A capacitor used in conjunction with one or more components to remove ac.

Frequency. The number of complete hertz that occur.

Frequency conversion. The process of changing the frequency of a signal by combining it with another frequency of a different value.

Frequency drift. A gradual change in frequency from a specified value. Prevalent in oscillator stages; usually caused by temperature changes within components.

Frequency modulation (fm). The process of varying the transmitter carrier above and below its normal resting frequency in accordance with the voice signal.

Frequency stability. The ability of an oscillator circuit or other a-c producing device to maintain a given frequency. Deviation limits from this frequency are usually given in percentage.

Frequency tolerance. The maximum frequency variation permissible on either side of the carrier frequency.

Generator. A mechanical or electronic device capable of producing a voltage. For example, an automobile generator converts mechanical energy into electrical energy.

Grounded. Connection to earth, or to a conductor that simulates earth.

Ground-plane antenna. A type of vertical antenna with radials extending from the base, and having a low angle of radiation so practically all of the r-f energy is confined to ground waves.

Ground waves. Radio waves that travel principally along the earth's surface.

Half-wave antenna. An antenna with the main element(s) equal in physical length to one-half the wavelength of the received or transmitted frequency.

Harmonic. An integral multiple of any frequency. For example, the frequency of a second harmonic is twice that of its fundamental.

Hertz. Cycles per second.

Heterodyne frequency meter. An instrument for determining the frequency of a transmitter or other signal-producing device.

Horizontally polarized wave. An electromagnetic wave whose electric lines of force are parallel to the earth's surface.

Impedance. The total opposition offered to the flow of alternating current.

Impedance match. The condition whereby the impedance of a circuit or component is equal to the impedance of its source.

Inductive coupling. A method by which energy from a coil is induced into another coil.

Input capacity. The sum of all direct capacitances at the input of a circuit.

Interference. Any noise or disturbance that affects the reception of radio signals.

Intermediate frequency (if). The difference frequency produced by mixing the incoming rf and the local-oscillator signals.

Intermediate-frequency amplifier. The amplifier that boosts the i-f signal.

Inverter. A device that changes dc into ac.

Kennelly-Heaviside layers. Layers of ionized gases between 50 and 400 miles above the earth. These layers tend to deflect the radio signals back to earth and thus make long-distance communications possible.

Kilohertz (kHz). 1000 hertz.

Kilowatt (kw). 1000 watts.

Lead-in insulator. A tube inserted through a hole drilled in a wall or window, through which the antenna lead-in wire is run.

Leakage. Undesirable current flow over or through an insulating material.
Line drop. The amount of line voltage lost between the source and the equipment due to leakage, reactance, and d-c resistance in the line.
Line filter. A network connected between the voltage source and the radio equipment to remove unwanted noise signals.
Line voltage. The a-c voltage at the wall receptacle.
Load. A component or device placed into a circuit to absorb power so it can serve a useful purpose. Also, the amount of power taken from a circuit.
Loading coil. A device used on an antenna to make it electrically longer than it is physically.
Matching. Connecting two circuits or components together so their impedances are equal.
Megahertz (MHz). 1,000,000 hertz.
Milliampere. 1/1000th of an ampere.
Milliwatt. 1/1000th of a watt.
Mixer. The stage where two different signal frequencies are combined to form a third frequency.
Modulate. To vary the amplitude or frequency of a signal in accordance with another.
Neutralization. Any process that balances out or prevents an undesirable effect.
Noise limiter. A network or circuit used to reduce or eliminate noise pulses.
Ohm's law formula. The formula that expresses the relationship between voltage, current, and resistance in a circuit—namely, $E = I \times R$.
Oscillator. A vacuum tube or transistor stage capable of generating an a-c signal.
Output stage. The final stage in the radio equipment.
Overtone. A harmonic of a fundamental frequency.
Parallel-resonant circuit. A tuned circuit whose inductance and capacitance are connected in parallel.
Piezoelectric. A term meaning "pressure electricity." Quartz, for example, will produce a small voltage when force is applied to it, or vice versa.
Pi-network. A network consisting of three impedances connected to resemble the Greek letter π.
Plate. The electron-collecting electrode of a vacuum tube. Also called the anode.
Plate current. The current which reaches the plate in a vacuum tube.
Plate modulation. A process whereby the signal in the final r-f transmitter stage is varied in step with a modulating signal injected into the plate circuit.
Plate power input. The plate voltage times plate current of the final r-f transmitter stage, measured with no modulation present.
Potentiometer. A device having a circular resistance element with a rotating contact arm that can be set at any desired point along its surface.
Power transformer. An iron-core transformer that usually has a number of secondaries to provide a variety of a-c voltages.
Push-pull circuit. An amplifier circuit containing two tubes, each conducting during one-half of the cycle, so that their combined output is more than that of a single tube.
Quartz. A mineral having piezoelectric properties, commonly used as the frequency-determining element (crystal) in oscillator circuits.
Radiate. To emit energy, such as radio waves, into free space.
Radio frequency (rf). The frequencies at which useful electromagnetic radiation can be obtained for communication purposes.
Radio-frequency amplifier. The stage that boosts r-f frequencies.
Radio spectrum. The area between the highest and lowest frequencies capable of producing usable radio waves.
Radio wave propagation. The diffusion of electromagnetic waves into space.
Rectifier. A device through which current can flow in one direction only.
Refracted wave. A radio wave that has been bent from its original course.
Resistance. The opposition a device or material offers to the flow of direct current.
Resonance. The condition in which the inductance and capacitance values of a circuit are such that maximum response is obtained.
Selectivity. The ability of a receiver to select and reproduce signals from a desired station while rejecting signals from other stations on adjacent channels.

Sensitivity. The receiver's ability to reproduce weak signals so they can be heard with satisfactory volume.

Signal generator. A test instrument capable of producing an r-f signal of a known frequency.

Signal-to-noise ratio. The ratio of the intensity of a desired signal, to the intensity of existing noise signals.

Single-ended stage. A stage having only one tube, as opposed to a push-pull stage.

Sky waves. Waves that travel into the upper atmosphere, and may or may not be reflected back to earth.

Squelch. A circuit that automatically mutes the receiver when no signal is being received, or when the signal is below a predetermined level.

Static. A cracking or popping type of interference caused when static electricity discharges.

Superheterodyne receiver. A receiver in which the incoming carrier signal is mixed (heterodyned) with another signal to produce an intermediate-frequency carrier.

Superregenerative detector. A detector circuit in which part of the output signal is fed to the input and regenerated.

Suppressor. A device used primarily in vehicle installation to reduce noise-pulse interference.

Temperature coefficient. The change in characteristics of a substance for each degree centigrade change in temperature.

Thermistor. A resistor whose value varies with temperature, used to compensate for the effects of temperature variations on circuit operation.

Transceiver. A transmitter and receiver mounted on a single chassis and having one or more stages common to both sections.

Transmission line. Any conductor used to convey signal energy from one point to another.

Trimmer capacitor. A small adjustable capacitor used principally in the tuning circuits of radio equipment.

Ultra-high frequency (uhf). Any frequency within the 300- to 3000-MHz range.

Universal power supply. A power supply designed to operate from one or more ac and/or dc voltages.

Vertical antenna. A nondirectional antenna that operates from a perpendicular position.

Vertically polarized wave. An electromagnetic wave whose electric lines of force are perpendicular to the earth's surface.

Vibrator. An electromagnetic device which, by means of vibrating contacts, chops up direct current in the primary circuit of a transformer to provide ac across the secondary.

Voice coil. A coil, attached to the cone of a dynamic speaker, which moves as the audio signal passes through it.

Voltage doubler. A rectifier circuit that doubles the voltage it receives.

Voltage drop. The voltage developed across a component by the current flowing through it.

Wavelength. The distance a radio wave travels during one cycle.

Zero beat. The condition in which a beat frequency cannot be produced because the two frequencies being combined are of the same value.

Index